Table of Contents

Executive Summary 1

1 Introduction 1-1
   1.1   Scientific Introduction 1-1
       1.1.1   The Diversity of Exoplanets 1-1
       1.1.2   Exoplanet Atmospheres... 1-3
       1.1.3   An Anticipated Diversity for Planet Habitability and Biosignature Gases.......... 1-4
       1.1.4   Why Space-Based Direct Imaging? 1-5
       1.1.5   Summary 1-5
   1.2   Technical Introduction 1-5
       1.2.1   Starshade Conceptual Introduction 1-5
       1.2.2   History 1-5
       1.2.3   Starshade Strengths.. 1-6
       1.2.4   Programmatic Challenges 1-8
       1.2.5   Technical Challenges 1-8
       1.2.6   Summary 1-9
   1.3   The Exoplanet Science Landscape in 2024 1-9
       1.3.1   Indirect Detections Using Stellar Reflex Motion.................................... 1-9
       1.3.2   Transits 1-10
       1.3.3   Exoplanet Imaging Detections 1-10
       1.3.4   Disk Imaging 1-11
       1.3.5   Summary 1-12

2 Science Goals and Objectives 2-1
   2.1   Science Goals 2-1
   2.2   Detailed Description of Science Objectives... 2-1
       2.2.1   Identifying Exoplanets... 2-1
       2.2.2   Characterizing Exoplanets via Spectroscopy.. 2-2
       2.2.3   Known Giant Planets 2-5
       2.2.4   Circumstellar Disk Science 2-6
   2.3   Astrophysical Contaminants... 2-8
       2.3.1   Exozodiacal Clumps and Background Objects 2-8
       2.3.2   Scattered Light from Companion Stars.. 2-9
       2.3.3   Background Stars 2-10
       2.3.4   Extragalactic Sources 2-10
       2.3.5   Summary of Observing Protocols 2-11
       2.3.6   Summary of Preparatory Science Recommendations 2-12

3 Design Reference Mission 3-1
   3.1   Key System Performance Parameters 3-1
   3.2   Integration Times 3-3
   3.3   Target Lists 3-4
       3.3.1   Candidate Earth Twins 3-4
       3.3.2   Known RV Planets 3-5
       3.3.3   Jupiter Twin Survey 3-5
   3.4   Observing Sequence... 3-6
   3.5   Summary.. 3-7

Executive Summary

Can we find another planet like Earth orbiting a nearby star? To find such a planet would complete the revolution, started by Copernicus nearly 500 years ago, that displaced Earth as the center of the universe.... Astronomers are now ready to embark on the next stage in the quest for life beyond the Solar System—to search for nearby, habitable, rocky or terrestrial planets with liquid water and oxygen.... The observational challenge is great, but armed with new technologies...astronomers are poised to rise to it.
–New Worlds, New Horizons, 2010

Figure ES-1. To the rest of the Universe, Earth appears as an exoplanet. The starshade-telescope mission is capable of imaging exoplanets with the properties of Earth around 20 target stars. Image: Earth as seen from the Voyager I spacecraft at a distance of 4 billion miles.

For the first time in human history, the technological reach exists to discover and characterize planets like Earth orbiting stars other than the Sun. A space-based direct imaging mission to ultimately find and characterize other Earths is a long-term priority for space astrophysics (NRC 2010).

The Exo-Starshade (Exo-S) Science and Technology Definition Team (STDT) is tasked by NASA to study the starshade-telescope mission concept under the "Probe" class of space missions, with a total cost of less than $1B (FY15 dollars). Per the STDT charter, the mission should be ready for a "new start" in 2017, with launch in 2024, and the science must be beyond the expected ground capability at the end of the mission. The Exo-S mission concept study began in May 2013 and will run until the Final Report delivery in January 2015.

Science Goals and Program

Exo-S is a direct imaging space-based mission to discover and characterize exoplanets. With its modest size, Exo-S bridges the gap between census missions like Kepler and a future space-based flagship direct imaging exoplanet mission. With the ability to reach down to Earth-size planets in the habitable zones of nearly two dozen stars, Exo-S is a powerful first step in the search for Earth-like planets with atmospheric biosignature gases: compelling science can be returned at the same time as the technological and scientific framework is developed for a larger flagship mission. The Exo-S mission has four science goals.

The first goal is *to discover planets, from Jupiter size down to Earth size, orbiting nearby Sun-like stars*. Within this discovery goal is the possibility to find Earth-size exoplanets in the habitable zones of about 20 Sun-like stars— arguably one of the most exciting pursuits in exoplanet research (Figure ES-1).

The second science goal is *to measure spectra of a subset of newly discovered planets*. The Exo-S spectral range is from 400– 1,000 nm, with a spectral resolution of $R = 70$. Of particular interest are the so-called sub-Neptunes, planets 1.75 to 3 times the size of Earth that are very low density, with no solar system counterparts and unknown composition. Spectral resolution for Earth-size planets will depend on the target brightness.

The third science goal is designed to guarantee outstanding science return: *to measure spectra of known giant planets*, detectable by virtue of extrapolated position in the 2024 timeframe. Molecular composition and presence of clouds or haze will yield information on the diversity of giant planet atmospheres. For the known Jupiters with radial velocity orbits, the first observation will yield the planet orbital inclination and therefore the planet mass.

The fourth science goal is *to study circumstellar dust*. Observations will shed light on the dust-generating parent bodies (asteroids and comets), as well as possibly point to unseen planets below the mission's direct detection thresholds.

Direct imaging exoplanet science is a daunting task not afforded justice by a few outlined goals. Several pressing astrophysical questions have come to the forefront, including how much can be learned about planets with limited spectral and temporal information, how planets can be efficiently distinguished from background sources, how stray light from binary stars should be handled, and how higher exozodiacal dust levels than the solar system's might impact the science harvest of a direct imaging mission. Answering these concerns will require a large-scale dedicated effort in the coming years.

Observing Program

An observing strategy is created from balancing the search for new exoplanets with the spectral characterization of known Jupiters, while factoring in the time it takes to align the starshade and telescope system to observe the next target star, and therefore the number of possible retargets available within the mission lifetime. The star list includes a total 53 target stars with 20 searchable for candidate exo-Earths; 17 target stars with 19 known giant planets at favorable elongation for spectral characterization; and 16 additional "Jupiter search" target stars. Two of these targets also have known debris disks. The example observing plan is to observe each target star once during the first 22 months, with follow-on observations scheduled for the rest of the mission lifetime (14 months), for confirmation of potential detections and spectroscopic observations. The actual observing schedule is adaptable to real-time discoveries.

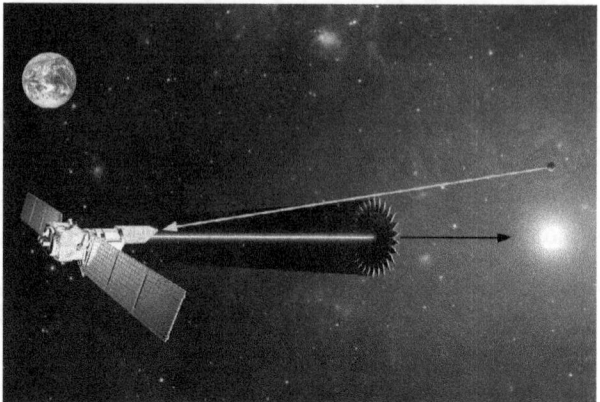

Figure ES-2. A starshade, also called an external occulter, is a precisely shaped screen that flies in formation with a telescope. The starshade blocks starlight to create a high-contrast shadow so that only planet light enters the telescope.

Starshade Description and Unique Advantages

A starshade flies in formation with a telescope and employs a precisely shaped screen, or external occulter, to block starlight, creating a high-contrast shadow that enables direct imaging of planets (Figure ES-2). Most designs feature a starshade tens of meters in diameter that is separated from the telescope by tens of thousands of kilometers.

The main strength of a starshade mission is that the starshade itself is nearly completely responsible for starlight suppression. Most significantly, the inner working angle (IWA, the closest angle on the sky at which a planet can be imaged) and the planet-star flux contrast achieved in the telescope image (the reduction in starlight at the planet location) are both independent of the telescope aperture size. This "decoupling" of the IWA from telescope diameter enables detection of planets down to the size of Earths with a small and relatively simple space telescope.

The starshade mission drawback is the length of time needed to realign the starshade and telescope for each new target star, which can take from several days to a couple of weeks. Nevertheless, multiple feasibility studies performed over the last several years demonstrate that a compelling search and characterization program is achievable.

A starshade offers many additional critical advantages including: unlimited outer working angle (OWA) for outer planet observing; broadband visible to near-infrared wavelength operation; and high throughput for efficient spectroscopy. Additionally, and most importantly, no special optical or wavefront control requirements are imposed on the telescope because the starshade itself performs the starlight suppression.

Baseline Mission Design

A viable and compelling starshade-telescope mission for exoplanet direct imaging is possible within the $1B cost constraint, incorporating cost-driven specific design solutions for several mission aspects.

The starshade and telescope share the same low-cost launch vehicle, conserving cost (Figure ES-3). An Earth-leading orbit is the slightly preferred option for a dedicated exoplanet mission over an Earth-Sun L2 orbit because the negligible gravity gradient in an Earth-leading orbit requires less fuel for starshade-telescope formation flying.

The telescope, instrument, and spacecraft bus systems are kept to low-cost units with extensive flight heritage. The telescope is a 1.1-m-diameter aperture commercially available telescope used for Earth imaging (NextView; with four currently operational),

with the predominant modification being the addition of a lightweight sunshade. The telescope has a conventional instrument package that includes the planet camera, a basic spectrometer, and a guide camera.

The telescope bus is a heritage design from an analogous mission: the European Space Agency's (ESA) PROBA-3 (PRoject for OnBoard Autonomy) mission, scheduled for launch in 2017. Key but straightforward modifications are needed for the increased mass and more distant orbit of the starshade-telescope system as compared to the PROBA-3 mission. The starshade bus will be a simplified version of the telescope bus.

The starshade has not been flown before, but extensive heritage from large deployable antennas with comparable development risk makes the starshade development manageable. Details of the starshade mechanical and optical design, fabrication, deployment, and technology development are provided in this report. The fully deployed starshade will be 34 m in diameter, consisting of a 20-m inner disk and 28 petals of 7 m length (Figure ES-4). The specific shape of the petals is found via an optimization process that creates the possible broadband shadow at the telescope aperture for a given starshade diameter and petal length. A perimeter truss, with deployable antenna heritage, forms the inner disk and controls deployed petal positions.

Technology Development

Full-scale, ground-based end-to-end testing is not possible for the full starshade-telescope system; rather, it is replaced by a two-step process. First, subscale testing will demonstrate

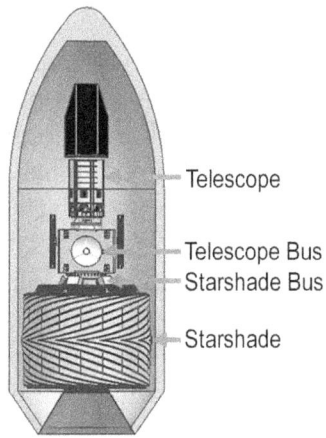

Figure ES-3. Baseline launch configuration of the telescope and starshade system shown in a 5-m launch fairing. The starshade will stow compactly around a load-carrying hub structure.

- Telescope
- Telescope Bus
- Starshade Bus
- Starshade

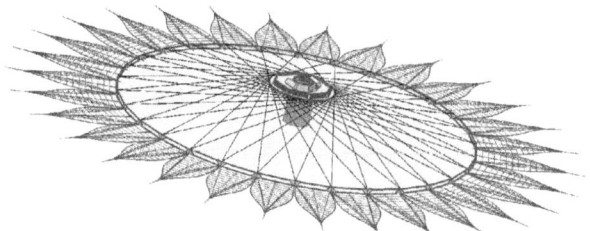

Figure ES-4. Fully deployed 34-meter starshade.

a dark shadow in broadband light in the lab and will validate the optical model to the required levels of a few times 10^{-11} contrast. The dark hole formed by a full-size starshade at its distance of 37,000 km is described by exactly the same diffraction equations as a small-scale starshade in a laboratory facility. To date, laboratory testing of starshades several centimeters in diameter has validated optical propagation models by achieving starlight suppression in monochromatic light to a few parts in 10^{-10}, close to required flight levels.

Second, metrology tests of the full-scale flight starshade will verify that the starshade will have the correct shape on-orbit. A precision manufactured petal prototype has demonstrated that a starshade petal can be manufactured to the required shape tolerances with flight-like materials. Deployment tests have shown that the petals can be deployed to the required position tolerances (Figure ES-5). The testing program gives high confidence that a properly constructed starshade will perform as predicted on-orbit.

Beyond optical model validation, precision deployment, and shape control, the remaining starshade engineering challenges (primarily related to long-distance formation flying and stray light control) are well understood and achievable (see Section 6).

The Starshade as a "Stand-Alone" Option to Rendezvous with an Existing Space Telescope

The option to make the National

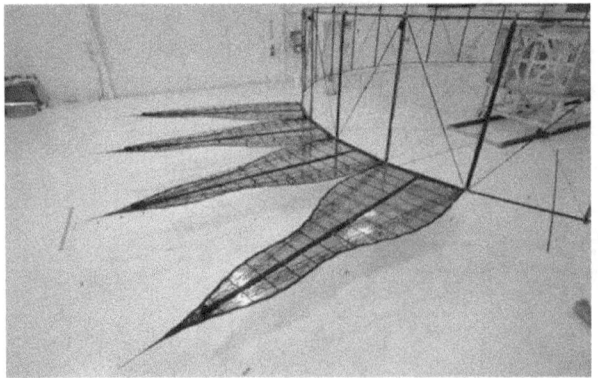

Figure ES-5. Deployment demonstration of a partial starshade prototype from 2013 at Northrop Grumman Corporation facilities.

Reconnaissance Office (NRO) WFIRST-AFTA (Wide-Field Infrared Survey Telescope–Astrophysics Focused Telescope Assets) and other space telescopes "starshade ready" with a starshade launched separately is an attractive alternative to the baseline mission design. This option will be explored during the second part of the Exo-S STDT study. A starshade-ready telescope would have its spacecraft bus outfitted before launch to include a communication system and a small guide camera integrated with the exoplanet instrument, so that the telescope can operate in formation with the starshade. The starshade can be constructed and launched separately, at a later time than the telescope, and could rendezvous with the space telescope in orbit, either after the space telescope's prime mission, or during the prime mission as a secondary experiment operating, for example, only 20% of the time.

The WFIRST-AFTA, or a similarly sized telescope, offers a huge improvement in exoplanet imaging capability when operated with a starshade. The larger telescope offers an order of magnitude reduction or better in integration times relative to the baseline mission. This translates into an increase in the number of target stars observed and increases the spectral resolution at which Earth-size planets can be characterized to the full desired level of $R = 70$.

Summary

The starshade-telescope system probe-class mission offers a breakthrough opportunity for space-based exoplanet direct imaging: it is the only way to reach well into the habitable zones of nearby stars to detect and characterize Earth-sized exoplanets using a relatively small space telescope. This capability is due to the planet-star flux contrast and IWA being nearly independent from the telescope aperture size. The starshade is responsible for blocking the starlight, enabling a non-specialized space telescope. Starshade technology progress is on track for a new start in 2017.

1   Introduction

1.1   Scientific Introduction

We stand on a great threshold in the human history of space exploration. On the one side of this threshold, we know with certainty that planets orbiting stars other than the Sun exist and are common. These worlds beyond our solar system are called exoplanets, and astronomers have found (statistically speaking) that every star in our Milky Way Galaxy has at least one planet (Cassan et al. 2012). On the other side of this great threshold lies the robust identification of Earth-like exoplanets with habitable conditions, and with signs of life inferred by the detection of "biosignature gases" in exoplanetary atmospheres. To bridge this divide, a space-based telescope, above the blurring effects of Earth's atmosphere, is needed to block out the starlight and search for planets directly. In this way, small rocky planets (planets with sizes less than 1.75 Earth radii) orbiting nearby Sun-like stars can be found, some with the potential to host life.

The goal of the Exo-Starshade (Exo-S) Science and Technology Definition Team (STDT) is to demonstrate a viable starshade-telescope space mission concept under $1B that has a compelling and impactful direct imaging exoplanet science program. The starshade-telescope system has the goal of searching for new exoplanets, from Jupiter size down to Earth size, including the capability to discover rocky exoplanets in the habitable zones (HZs) around a subset of favorable target stars. A second major mission goal is to spectroscopically characterize both new and already known exoplanets. Circumstellar dust will be observed and will help understand exozodiacal dust levels and the dust-generating parent bodies (asteroids and comets), as well as possibly point to unseen planets below the mission's direct detection thresholds.

1.1.1   The Diversity of Exoplanets

Central to the foundation of the astronomical search for exoplanets is the frequency, or occurrence rate, of various planet types. The commonality and diversity of exoplanets is remarkable. Of most excitement to the public and exoplanet community is the NASA Kepler space telescope findings that small planets are extremely common in our galaxy (Figure 1.1-1). Kepler has specifically found that: planets 1.75 to 3 times the size of Earth

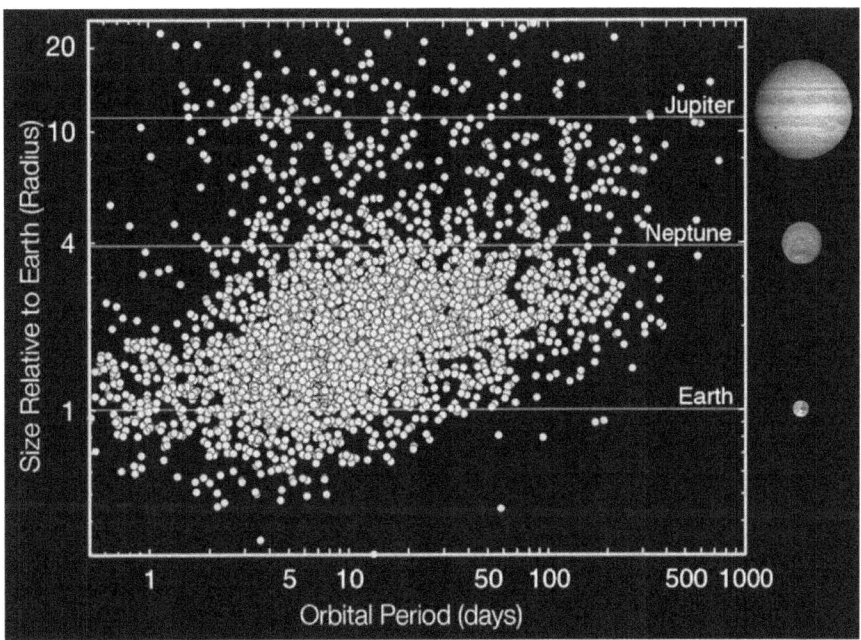

Figure 1.1-1. Kepler's planets and planet candidates as reported in 2013. Credit: NASA/Kepler mission.

are nearly 10 times more common than giant planets (Fressin et al. 2013 and Howard 2013; Figure 1.1-2); multiple planet systems somewhat reminiscent of our inner solar

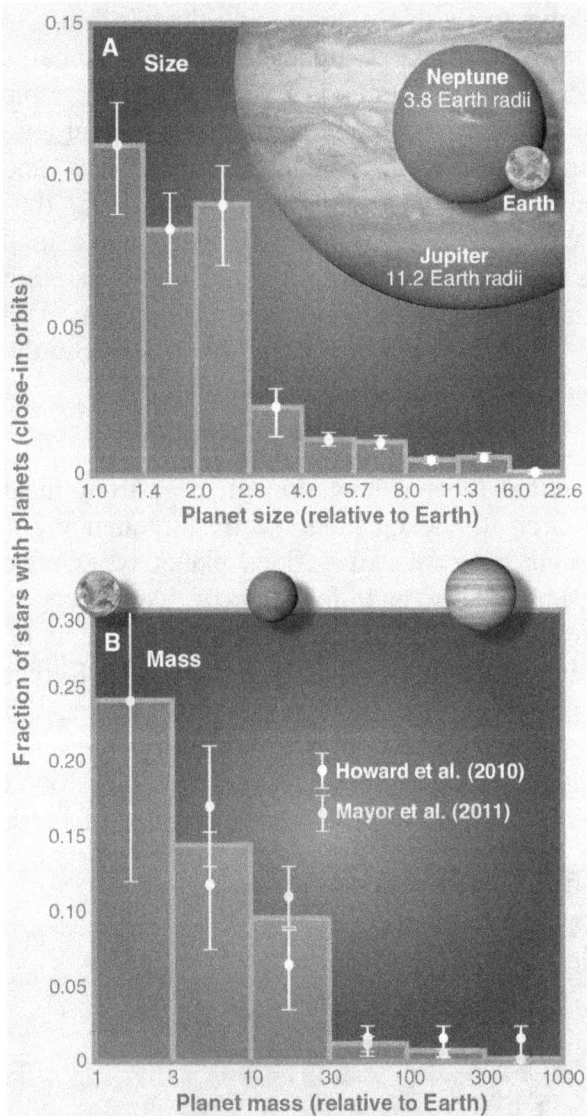

Figure 1.1-2. The (A) size and (B) mass distributions of planets orbiting close to G- and K-type stars.

The distributions rise substantially with decreasing size and mass, indicating that small planets are more common than large ones. Planets smaller than 2.8 R_E or less massive than 30 M_E are found within 0.25 AU of 30 to 50% of Sun-like stars. (A) The size distribution from transiting planets shows occurrence versus planet radius. (B) The mass (Msini) distributions show the fraction of stars having at least one planet with an orbital period shorter than 50 days (orbiting inside of ~0.25 AU). Both distributions are corrected for survey incompleteness for small/low-mass planets to show the true occurrence of planets in nature. (Image and caption from Howard 2013.)

system are common (Lissauer et al. 2014; Rowe et al. 2014); and approximately 1 in 5 Sun-like stars have an Earth-size planet in the star's habitable zone (Petigura et al. 2013).

While most of the Kepler results are for planets with orbital periods of 200 days or less, due to observational selection effects, a logical extrapolation is accepted as a solid inference: that rocky planets in habitable zones are common enough to motivate a space mission that can sample a couple dozen stars for rocky planets.

From a variety of observing techniques, the point that planets exist at all masses, semi-major axes, and orbits is well established, motivating that there will be a variety of interesting planets to discover in any parameter space (Figure 1.1-3).

The commonality of small exoplanets includes a completely unexpected finding: the existence of planets with no solar system counterparts, the so-called sub-Neptunes of > 1.75 to 3 Earth radii. The prevalence of sub-Neptunes is supported for a variety of orbital separations by the transit technique (Fressin et al. 2013 and Howard 2013), radial velocity (RV) surveys (Howard 2013), and microlensing studies (Sumi et al. 2010). Planets are referred to as sub-Neptunes if they have a "gas envelope" that is far more substantial than any atmosphere, and super Earths or rocky planets if they are predominantly rocky with thin atmospheres. The difference is a critical one—planets with thick atmospheres will be too hot at any surface to accommodate life.

The sub-Neptunes are a boon for direct imaging searches because their large size (larger than Earth) make them more easily detectable. Moreover, atmosphere observations may be the only way to shed light on their very nature—their actual atmospheric and interior composition is not known. Most of the sub-Neptunes have surprisingly low mean densities. The sub-Neptunes may be so-called "water worlds" (with 50% or more water by

Photo Removed Due to Copyright Restrictions

Figure 1.1-3. Known exoplanets with measured masses or minimum masses. Data from NASA Exoplanet Archive (image courtesy of P. Lawson).

mass with thick steam atmospheres), or massive rocky planets that have significant H or H/He envelopes, or smaller versions of Neptune that have a water or methane interior with significant H or H/He envelopes (see Figure 2.2-7, and Rogers and Seager 2010).

The field of exoplanets has seen many other revolutionary discoveries in the last decade, all supporting the diversity of exoplanets and exoplanetary systems. Most of the findings are related to uncovering new

populations of exoplanets and defining their characteristics, for example: hot super Earths (Batalha et al. 2011); circumbinary planets (Doyle et al. 2011); compact multiple planet systems; planets with a suggested high carbon-to-oxygen (C/O) ratio (dubbed "carbon planets"; Madhusudhan et al. 2011); and others. Though not all are related to direct imaging planet discovery missions, the vast array of findings supports the sentiment that the discovery space for exoplanets is large.

1.1.2 Exoplanet Atmospheres

The diversity of exoplanets is expected to extend to planet atmospheres. Out of dozens of exoplanet atmosphere observations (Seager and Deming 2010; Madhusudhan et al. 2014), a handful of hot transiting exoplanets have detailed atmosphere measurements across a wide wavelength range. The example of the transiting hot Jupiter observed in transmission with space and ground-based telescopes is a good one because it is surprisingly dominated by haze, unexpected because the atmosphere was previously thought to be too hot to support haze (Figure 1.1-4). The hot transiting planet spectra are enough to gather a glimpse that planets that are similar in size, mass, and parent star type have different atmospheres. On the other hand, a few hot Jupiters observed via secondary eclipse thermal emission at 2 μm show water vapor as expected (measured by

Photo Removed Due to Copyright Restrictions

Figure 1.1-4. Transit transmission spectra of the hot Jupiter exoplanet HD 189733b with data points from HST STIS, ACS, WFC3, NICMOS, and Spitzer. The grey line shows a synthetic spectrum with a dust-free model. The dotted lines, from left to right, indicate the effect of Rayleigh scattering at 2000 K, 1300 K, a cloud with grain sizes increasing linearly with pressure and an opaque cloud deck. HD 189733 b surprised the community with its presence of haze and/or clouds. (From Pont et al. 2013.)

the Hubble Space Telescope [HST] Wide Field Camera 3 [WFP3]; Deming et al. 2013 and Wakeford et al. 2013).

Atmosphere observations have so far been limited to hot transiting exoplanets, observed either in transmission or secondary eclipse thermal emission. A few massive or young giant planets at very large orbital separations from the star have atmospheres observed from ground-based telescopes in near-infrared (NIR) narrow bands (J, H, and K).

The prospect of obtaining spectra of over one dozen known giant planets with planet-star orbital separations from 1 to a few AU will enable comparison of giant planets of very different kinds.

1.1.3 An Anticipated Diversity for Planet Habitability and Biosignature Gases

The variety of exoplanets in terms of orbits, masses, and possibly atmospheres, is now established. As a consequence, habitable planets may vary widely and be different from the Earth analog (see Figure 1.1-5 and Seager 2013 and references therein). The fundamental reason is that surface temperatures are governed by the atmospheric greenhouse properties and the range of atmospheric composition and mass is not predictable *a priori*.

For example, it is possible a 10 Earth mass, 1.75 Earth radii planet with an H_2-dominated atmosphere could be habitable and host biosignature gases (Seager et al. 2013).

While the mission design shouldn't be constrained by all theoretically considered habitable planets (for further examples see Domagal-Goldman et al. 2011 and references therein), the opportunity that planets larger than Earth could host life must be acknowledged since the difference between a planet of 1.75 and 1 Earth radii is significant from the standpoint of observational detection.

Most of the gases in the Earth's atmosphere that exist to the 100 parts per

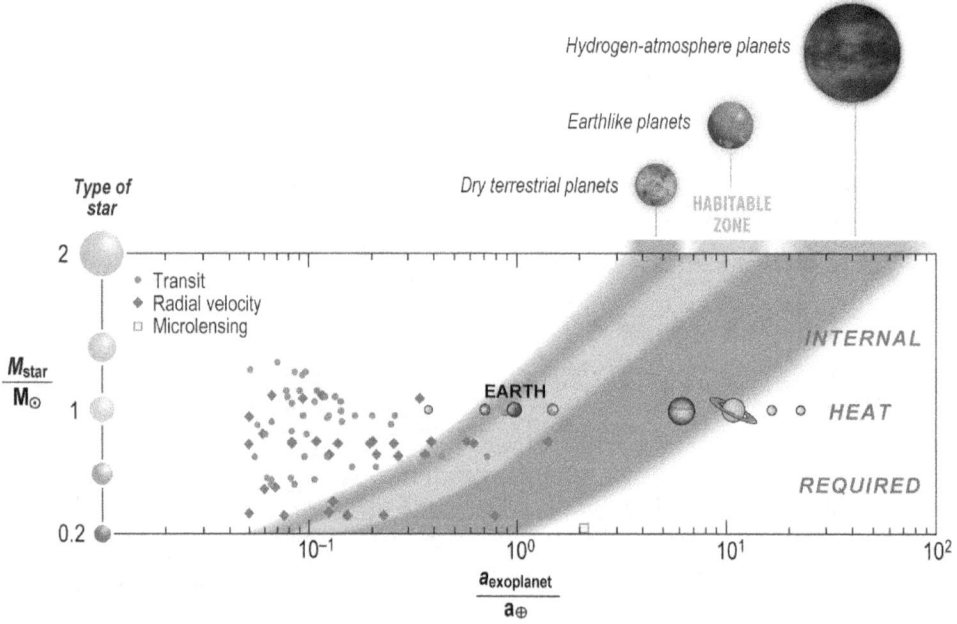

Figure 1.1-5. The extended habitable zone. The light blue region depicts the "conventional" habitable zone for N_2-CO_2-H_2O atmospheres. The yellow region shows the habitable zone as extended inward for dry planets, as dry as 1% relative humidity. The outer purple region shows the outer extension of the habitable zone for hydrogen-rich atmospheres and can even extend out to free floating planets with no host star. The solar system planets are shown with images. Known super Earths (here planets with a mass or minimum mass less than 10 Earth masses) taken from Rein 2012. (From Seager 2013.) See Seager 2013 and references therein. For a discussion of the inner edge of the habitable zone see Zsom et al. 2013 and references therein. For a discussion of the traditional habitable zone, see Kopparapu et al. 2013.

trillion level, with the exception of the noble gases, are produced by life, although most of them are also naturally occurring. Therefore, while oxygen is touted as the most robust biosignature gas, the need for a broad spectral range is essential. The plan for biosignature gases is to be prepared to detect gases that are many orders of magnitude out of chemical equilibrium, and eventually identify them as being produced by life.

1.1.4 Why Space-Based Direct Imaging?

The space-based direct imaging search for Earths is a natural and essential next step in a continuing series of NASA exoplanet missions. Only space-based direct imaging can eventually find and identify true Earth analogs, and study atmospheres of small planets orbiting Sun-like stars.

Although NASA's pioneering Kepler telescope discovered over 800 exoplanets and 2,500 more exoplanet candidates, the Kepler exoplanets are too distant from Earth for near-future follow-up studies of their atmospheres. While NASA's Transiting Exoplanet Survey Satellite (TESS; launch 2017; http://space.mit.edu/TESS/TESS/TESS_Overv iew.html) will perform an all-sky survey to find thousands of exoplanets orbiting nearby stars, the TESS rocky worlds accessible for atmosphere studies by the James Webb Space Telescope (JWST) are limited to small stars. Even if a future planet transit search mission, such as the ExoplanetSat small satellite constellation (Smith et al. 2010) or the European Space Agency's (ESA) Plato mission (Rauer et al. 2013), finds transiting Earths around nearby stars (keeping in mind the rarity of transiting planets at Earth's orbital separation: 1/200 if every star has an Earth), the atmospheres of such planets will be too thin to be observed via the transit transmission technique (Kaltenegger and Traub 2009).

To find small exoplanets bright enough for atmosphere characterization, including the eventual search for biosignature gases, we must find planets orbiting stars that are bright, i.e., close to our own Sun. There is a strong desire to know if small planets and planetary systems that resemble our own orbit our nearest neighboring stars.

1.1.5 Summary

The starshade-telescope system will impact exoplanet science in a foundational way by finding and characterizing exoplanets around the nearest Sun-like stars. "Comparative exoplanetology" by way of atmospheric spectroscopy will enable us to compare the sample of hot Jupiters to their colder Jupiter-type planet counterparts. Spectra of a number of sub-Neptunes has the potential to help us understand the nature of these enigmatic planets. If Earth analogs are common, the Exo-S mission has outstanding potential to detect planets of Earth size in the habitable zones of a couple dozen nearby Sun-like stars.

1.2 Technical Introduction

1.2.1 Starshade Conceptual Introduction

A starshade (also called an external occulter) is a spacecraft with a carefully shaped screen flown in formation with a telescope (Figure 1.2-1). The starshade size and shape, and the starshade-telescope separation, are designed so that the starshade casts a very dark and highly controlled equivalent of a shadow, suppressing the light from the star while leaving the planet's reflected light unaffected. In this way, only the exoplanet light enters the telescope. Most designs feature a starshade tens of meters in diameter that is separated from the telescope by tens of thousands of kilometers.

1.2.2 History

The idea of using an (apodized) starshade to image planets was first proposed in 1962 by Lyman Spitzer at Princeton (Spitzer 1962). In this landmark paper (in which he also suggested that NASA build and fly what later became the Hubble Space Telescope), he proposed that an external occulting disk could be used to block most of the starlight prior to

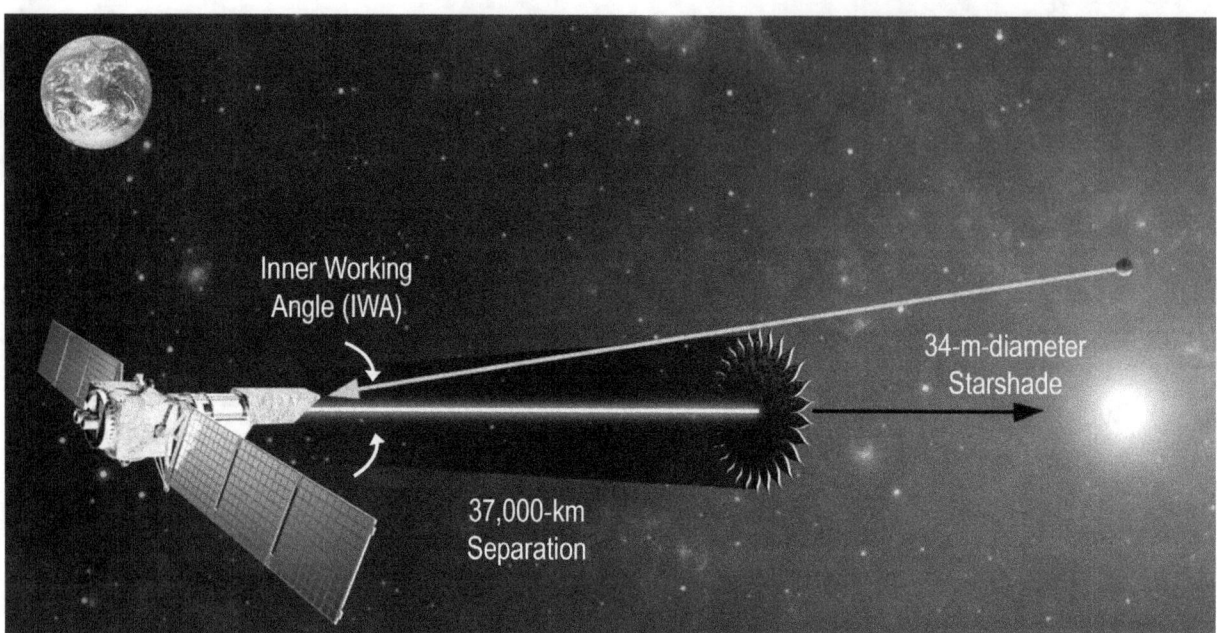

Figure 1.2-1. Schematic of the starshade-telescope system (not to scale). Starshade viewing geometry with inner working angle (IWA) independent of telescope size.

reaching the telescope pupil, thus enabling the direct imaging of planets around nearby stars. He realized that diffraction from a circular disk would be problematic for imaging an Earth-like planet due to an insufficient level of light suppression across the telescope's pupil. He posited that a different edge shape could be used instead, foreshadowing today's approach. In 1974, the idea was revived by G.R. Woodcock of the Goddard Space Flight Center using apodized starshades. In 1985, Marchal (1985) analyzed different analytic apodization functions for their suitability in a starshade starlight suppression system and suggested a starshade shaped remarkably like the flower petal one proposed here.

In the intervening years between 1985 and the present day, several mission concepts were proposed using apodized starshades. Most notably, Copi and Starkman in 2000 revisited the apodized starshade and found transmissive solutions defined by polynomials; their proposed mission was called the Big Occulting Steerable Satellite (BOSS). A few years later Schultz et al. (2003) proposed a similar mission dubbed UMBRAS (Umbral Missions Blocking Radiating Astronomical Sources).

However, these suggestions were hampered by the difficulty in manufacturing a transmissive surface within the tight tolerances necessary. In 2004, Simmons (2004 and 2005) first proposed using shaped-pupil approaches to replace an apodized starshade by a particular shape, in particular, the star-shaped mask (Vanderbei 2003) that became today's starshade solution. In his oft-cited *Nature* paper, Cash (2006) proposed using a hypergaussian function for the petal apodization. Shortly thereafter, Vanderbei et al. (2007) developed optimal starshade apodizations that minimize the starshade distance and size while achieving the desired contrast over wide spectral bands.

1.2.3 Starshade Strengths

There are several strengths that a starshade approach brings to exoplanet imaging and characterization. Most significantly, the inner working angle (IWA; the closest angle on the sky at which a planet can be imaged) and the contrast achieved in the telescope image (the reduction in starlight at the planet location) are both independent of the telescope aperture size. This differs from conventional coronagraphs where the area of high contrast in the image is a function of lambda/D, where D is the diameter of the telescope. With a starshade, the starlight is removed and the angle at which a planet is visible depends only on the size and distance of the starshade. In principle, even a tiny telescope would be adequate for direct imaging of small exoplanets. In practice, the telescope aperture must be sufficiently large because of the imperfect starlight cancellation of the starshade and the need for reasonable integration times.

The starshade, and not the telescope, is responsible for the achievable planet-star flux contrast. The fraction of starlight blocked by the starshade dictates the faintness of the planet that can be detected. Because the starlight never enters the telescope, there is no need for specialized optics to achieve high contrast (which typically reduce throughput), a relatively simple space telescope is all that is needed. On-axis obstructions or mirror segments do not interfere with starlight cancellation and wavefront correction is not required (which frees the telescope from tight thermo-mechanical requirements).

An additional significant feature of the starshade-telescope system is the absence of an outer working angle (OWA). A 360 degree suppressed field of view (FOV) with angles from the star limited only by the detector size is obtained with each image. This is particularly useful for imaging debris disks or planets at large orbital separations.

The starshade is also fundamentally broadband. Because all of the starlight suppression is created by a single optical element, a broad bandpass is obtainable. Furthermore, there are few constraints on other observatory astronomical instruments.

The starshade-telescope system can detect Earth-size planets in the habitable zone of Sun-like stars even with a small telescope (on order of 1-m aperture diameter). This ambitious statement is allowed by the fact that all of the starlight suppression is done by the starshade.

As long as the tolerances for starshade petal precision manufacturing, deployment, and formation flying control are met (see Table 1.2-1 and Section 6), the starshade will be capable of reaching the 10^{-10} contrast level needed to directly observe Earth analog exoplanets around Sun-like stars. An important related point supporting starshades with small telescopes is that wavefront correction is not required. If high-precision wavefront correction were required, the telescope collecting area would be a limiting factor on the starlight suppression, since wavefront sensing and control relies on collecting enough target starlight to sense the time-dependent optical imperfections that need to be corrected. In this case, small telescopes put Earth-Sun flux contrast levels out of reach.

The starshade's powerful capability for starlight cancellation means the challenges of reaching the required IWA and planet-star contrast all lie with the starshade. The challenges associated with producing a successful telescope-starshade system can be divided into "programmatic challenges" and "technical challenges".

Table 1.2-1. Summary of technology status and plans.

Key Challenges	Driving Spec	Technology Status
Dynamic stability	Deformations < 15 ppm after 10 s	Verified by analysis with large margins
Thermal stability	Non-uniform deformations ≤ 7.5 ppm	Verified by analysis with large margins
Manufacturing tolerance	Petal width < 100 μm (4 mil)	Demonstrated per TDEM-2009
Deployment tolerance	In-plane petal root position ≤ 0.5 mm	Demonstrated per TDEM-2010
Edge-scattered sunlight	Edge radius curvature <1 μm	Demo in progress per TDEM-2012
Laboratory contrast demo and model validation	10^{-10} contrast at flight Fresnel Number	Demo in progress per TDEM-2012
Formation flying	Sensing for lateral control ±1 m	Requires technology demonstration
Deploy control system	Avoid petal contact	Requires significant engineering effort
Launch restraint	Axial loads ≤ 10 g's	Requires significant engineering effort

1.2.4 Programmatic Challenges

Starshade programmatic challenges are innate limitations to the starshade-telescope system. First, a full-scale, ground-based end-to-end system test for the starshade-telescope system is impossible because of the large size of the occulting screen (tens of meters), the large separation distances between the telescope and starshade (tens of thousands of kilometers), and the guidance, navigation, and control (GN&C) formation flying requirements. Subscale lab (see Section 6.3) and field testing together with computer performance modeling and simulations are the only alternative.

The second programmatic challenge is operational: the starshade has a limited number of retarget maneuvers (on the order of 30 per year) due to retarget times (from several days to a couple of weeks), meaning that only a fixed number of stars can be observed over the mission duration. More than one starshade can mitigate the limited number of target stars.

1.2.5 Technical Challenges

The major technical challenges must be considered in light of flight-proven technologies for analogous commercial large deployable antenna systems in addition to highly successful starshade-specific NASA-funded technology demonstrations over the last several years. Technology readiness is detailed in Section 6 and this section provides a brief overview.

Key technology challenges, once considered tall-pole issues, but now considered demonstrated to be achievable are: precision petal manufacturing, precision deployed shape, and on-orbit stability.

Petals must be precisely manufactured to the specified petal width profile, or optical apodization function (width tolerance ≤ 100 μm). This capability was successfully demonstrated by a Technology Development for Exoplanet Missions (TDEM) activity (see Figure 1.2-2).

Petals must be precisely deployed to the specified petal root positions, as controlled by

the perimeter truss (in-plane root positions ≤ 500–750 μm). This capability was successfully demonstrated by a TDEM activity (see Figure 1.2-3).

Petal width profiles must be precisely maintained on-orbit (non-uniform thermal deformations ≤ 12 ppm). This capability was successfully demonstrated by analysis with large margins. Predicted deformations are a small fraction of allocations. Dynamic deformations are also allocated and successfully demonstrated by analysis with large margins, aided by the structural

Figure 1.2-2. Flight-like petal. See also Figure 6.2-1.

Photo Removed Due to Copyright Restrictions

Figure 1.2-3. Starshade stowage and deployment test with four petals. JPL/Princeton/NGC. See also Figure 6.2-3.

attenuation and damping provided by the starshade. Dynamic deformations are allocated after some transient period during which larger deformations are acceptable because they are not sensed by the instrument.

Key technology challenges, currently considered tall-pole issues and in work are: optical model validation, subscale contrast performance demonstration, and the control of edge-scattered sunlight. Activities are funded to address these issues prior to 2017.

One more open technology issue of lesser priority is formation flying at large separation distances. Formation flying precision is required to keep the telescope positioned within the dark shadow created by the starshade (lateral tolerance ≤ 1 m) and the separation distance within the range consistent with the optical bandpass (line of sight tolerance ≤ 250 km). The separation distance specification is very loose and is not actively controlled. Rather, corrections are applied as part of retargeting maneuvers. Mitigating the lateral control challenge are the very low disturbance forces afforded by the Earth-leading orbit. The formation flying challenge is primarily associated with sensing the starshade position. The baseline design accomplishes this with a fine guidance camera (FGC) operating with the telescope. An activity to demonstrate the sensing and control algorithms is proposed but not yet funded.

1.2.6 Summary

Starshade technology development has approached a point where successful technology demonstrations and well-defined technology gaps enable a clear path forward with manageable risk (see Figure 6-1, Technology Flow Chart). An appropriate funding effort for the remaining engineering challenges will enable achievement of technology readiness goals. For more details on the technology gap list and technology development plans, see Section 6.

1.3   The Exoplanet Science Landscape in 2024

Planetary systems consist of giant planets, sub-Neptunes, rocky (or terrestrial) planets, and belts of small bodies that generate debris particles. Ongoing research, upcoming developments in ground-based instrumentation, and the launch of new space missions promise to significantly advance our knowledge of these four exoplanetary system components in the coming decade. Nevertheless, a probe-scale exoplanet direct imaging mission can offer unique capabilities that will advance our knowledge of exoplanetary systems even further. Below is the likely context for exoplanet science at the time Exo-Coronagraph (Exo-C) / Exo-S would launch.

1.3.1 Indirect Detections Using Stellar Reflex Motion

RV surveys have detected almost 550 planets as of early 2014 (http://exocplanets.eu); the median orbital period of these detections is around 1 year. While the median semi-amplitude of these detections is 40 m/s (http://exoplanets.org; larger than the solar reflex velocity induced by Jupiter), only a dozen planets have measured RV semi-amplitude below 2 m/s. The smallest RV detection claimed to date has a 0.5 m/s semi-amplitude for the very bright star α Centauri B. Today's measurement precision of 50 cm/s is expected to improve toward 10 cm/s with the Very Large Telescope (VLT) Echelle SPectrograph for Rocky Exoplanet and Stable Spectroscopic Observations (ESPRESSO) and similar instruments on extremely large telescopes (European Extremely Large Telescope [E-ELT], Giant Magellan Telescope [GMT], Thirty Meter Telescope [TMT]). However, stellar RV jitter arising from star spots and activity sets a natural noise floor near 2 m/s (Bastien et al. 2014). Only in the quietest stars—or through careful averaging, filtering, and detrending of the data—will RV detections be achieved for semi-amplitudes below 1 m/s. By 2024, RV surveys should have detected any planets with periods < 20 yrs and with Saturn mass or greater around most bright stars, Neptune mass planets with period < 3 yrs around many stars, and 5 $M_\oplus$ super Earths with period < 1 yr in some systems. Complementary measurements of stellar astrometric wobble by the ESA Gaia all-sky survey will detect and measure orbit inclinations for planets of Jupiter mass or larger and periods < 5 yrs around unsaturated nearby stars (V > 6; Casertano et al. 2008). The orbital elements for the inner giant planets of nearby stars should be well in-hand by 2024.

1.3.2 Transits

Transit observations with the Kepler telescope (and with the CoRoT [COnvection ROtation and planetary Transits] telescope) has revealed the frequency and radius distribution of short-period (P < 1 yr) exoplanets by photometrically monitoring selected fields of solar-type stars. The 2017 TESS mission will identify shorter-period (P $\prec$ several weeks) planets around half a million bright field stars distributed around the sky. Around M stars, TESS detections will extend down to 1 $R_\oplus$ in the habitable zone.

Radial velocity follow up of TESS detections will reveal their mass distribution and the mass-radius relationship. Spectroscopic measurements made during transit and secondary eclipse by JWST, ELTs, and other facilities will constrain the temperatures and albedos of these planets, and for clear, low-molecular weight atmospheres, detect high-opacity atmospheric species such as Na I, H_2O, and CH_4. By 2024, transit work should have built a strong statistical picture of the bulk properties of inner planetary systems and led to atmospheric spectral information for many of their larger objects.

1.3.3 Exoplanet Imaging Detections

Only a handful of exoplanets have been directly imaged in their near-infrared thermal emission (e.g., Marois et al. 2010; http://exoplanets.eu). The small set of detections to date is due to the limited contrast capabilities of current instrumentation (see Figure 1.3-1), especially at small angular separations from a star. A new generation of high-contrast imagers based on extreme

Photo Removed Due to Copyright Restrictions

Figure 1.3-1. Direct imaging contrast capabilities of current and future instrumentation. (From Lawson 2013.)

adaptive optics systems, including the Gemini Planet Imager (GPI) and VLT Spectro-Polarimetric High-contrast Exoplanet Research (SPHERE), is now being deployed to large ground-based telescopes. Dozens of exoplanet imaging detections at 10^{-7} contrast and ~0.5″ separation should be achieved by these systems in the near-infrared, enabling detection and spectroscopy of thermal emission from warm (T > 200 K; very young or massive) gas giant planets. An appropriately designed ELT instrument in the 30-m class would be capable of such detections at even smaller IWA (~0.12″), but with only modestly better contrast. However, extreme adaptive optics systems are not currently baselined for ELT first generation instruments.

Ground-based, high-contrast imaging is limited by rapid wavefront changes arising from atmospheric turbulence. For a solar twin at 10 pc distance (H mag 3), a deformable mirror sized to create a ~0.5″ radius dark field cannot suppress the residual speckles to levels fainter than 10^{-7} of the central star brightness. This limit is defined by the available photons per subaperture in a reduced coherence time (Oppenheimer and Hinkley 2009, Table 2) and is nearly independent of telescope aperture size. To detect fainter objects, speckle averaging and subtraction methods must be employed. It is unclear how well this could be done, as the temporal behavior of residual atmospheric speckles at 10^{-7} contrast has never been characterized. Experience at less challenging contrast levels suggests that detections a factor of 10 below the raw contrast floor should be achievable. A planet-star flux contrast of 10^{-8} would enable detections of thermal emission from nine massive giant planets around nearby solar-type stars (Stapelfeldt 2006). It has been suggested that ELTs could detect planets in reflected light as small as 1 $R_\oplus$ at this contrast level, if they are present in the 0.1 AU radius habitable zones of bright nearby M dwarfs (Guyon and Martinache 2013). However, the required

stellar properties (V < 7 for sufficient guidestar photons, d < 8 pc to resolve the habitable zone with an ELT) results in a null target set.

JWST NIRCam (Near Infrared Camera) coronagraphy should be capable of detecting companions at contrasts of 10^{-6} at separations beyond 1.5 arcsec, capturing objects like our own Jupiter in 4.5-μm thermal emission if they are orbiting the nearest M stars. The uncertain luminosity evolution of young giant planets clouds the picture somewhat (Marley et al. 2007), but it appears that some of the more massive planets orbiting nearby (d < 20 pc) young (age < 1 Gyr), low-mass (M < 1.0 Msun) stars could be in view by 2024.

1.3.4   Disk Imaging

Imaging of protoplanetary disks is being revolutionized by ALMA (Atacama Large Millimeter/submillimeter Array), which will be able to resolve dynamical structures driven by protoplanets at angular resolutions approaching 0.01 arcsec. Protoplanetary disks in the nearest star-forming regions (d ~ 150 pc) are ideal ALMA targets, as their high optical depths give them high surface brightness in the submillimeter continuum. Debris disks are found around older main-sequence stars, with many nearby (d ~ 25 pc) examples. They are optically thin with a much lower dust content and much fainter submillimeter continuum emission; it will therefore be a challenge even for ALMA to resolve their detailed structure. ALMA will map a limited number of the brightest debris disks (Ld/Lstar > 10^{-4}) at 0.1 arcsec resolution. In addition to their exoplanet imaging capability, new adaptive optics coronagraphs now being deployed to large ground telescopes should image bright debris disks with comparable resolution and with sensitivity a few times better than ALMA but in the near-infrared. Similar instruments on ELTs would extend the resolution and IWAs of such studies to 10 and 30 milliarcsec respectively. With its 0.3-arcsec resolution at 20 μm, JWST will resolve warm dust emission around a sample of nearby A-type stars. New warm disks identified by the Wide-Field Infrared Survey Explorer (WISE) mission will be particularly important targets. A wealth of new data detailing the internal structure of bright circumstellar disks will have emerged by 2024, seeding a new theoretical understanding of disk structure, dynamics, and evolution.

1.3.5 Summary

While the advances described above will be remarkable scientific milestones, they fall well short of the goal of obtaining images and spectra of planetary systems like our own. The TESS mission will detect inner terrestrial planets transiting nearby cool stars, but their spectroscopic characterization will be challenging even using JWST. High-contrast imaging will detect and characterize warm giant planets, but not cool objects at 10^{-9} contrast like our own Jupiter and Saturn in their orbits around a solar-type star. Sharp images of dusty debris disks will be obtained, but only those with optical depths several hundred times that of our own asteroid and Kuiper belts. Radial velocity and astrometric surveys will have identified the majority of nearby stars hosting giant planets. What is currently missing from the 2024 exoplanetary science toolbox are space observatories that can study photons from cool planets (ranging from giants down to super Earths) and resolve tenuous dust disks around nearby stars like the Sun.

2 Science Goals and Objectives

2.1 Science Goals

> *Exo-S mission science goals:*
> 1. *Discover planets from Jupiter size down to Earth size orbiting nearby Sun-like stars*
> 2. *Measure spectra of a subset of newly discovered planets*
> 3. *Measure spectra of currently known giant planets*
> 4. *Study circumstellar dust*

The Exo-S mission has four science goals. The first goal is *to discover planets from Jupiter size down to Earth size orbiting nearby Sun-like stars*. Within this discovery goal is the possibility of discovering Earth-size exoplanets in the habitable zones (HZ) of about 20 Sun-like stars—arguably one of the most exciting pursuits in exoplanet research (Figure ES-1).

The second science goal is *to measure spectra of a subset of newly discovered planets*. The Exo-S spectral range is from 400–1,000 nm, with a spectral resolution of up to $R = 70$, which will enable detection of key spectral features. Of particular interest are the so-called sub-Neptunes, planets with no solar system counterparts, about 1.75 to 3 times the size of Earth. These planets have very low densities compared to Earth, yet their actual composition is not known.

The third science goal is designed to guarantee outstanding science return: *to measure spectra of currently known giant planets*, detectable by virtue of extrapolated position in the 2024 timeframe. Molecular composition and the presence/absence of clouds or hazes will inform us of the diversity of giant planet atmospheres.

The fourth science goal is *to study circumstellar dust*. Observations will shed light on the dust-generating parent bodies (asteroids and comets), and the dynamical history of the system, as well as possibly point to unseen planets below the mission's direct detection thresholds. Furthermore, these dust measurements will be important for future follow-up observations of terrestrial planets in habitable zones. The lack of knowledge of such dust levels inside the habitable zones of nearby stars is currently a major unknown affecting mission planning for future flagship mission concepts.

2.2 Detailed Description of Science Objectives

2.2.1 Identifying Exoplanets

The Exo-S mission has a prime goal to discover new exoplanets. The path to identifying an exoplanet is to first rule out the observed point of light as being a background astrophysical source, and second to estimate the kind of planet so that the follow-up characterization strategy can be optimized according to the science goals for spectroscopy.

The overall path for identifying worthy candidates for precious follow-up spectroscopic observations hinges on eliminating possibilities until there is high likelihood that the candidate is a particular type of planet of interest. Such a "planet validation" philosophy is parallel to the Kepler Space Telescope's planet hunting process. In Kepler's case, planet validation was motivated because follow-up radial velocity (RV) observations to formally confirm an exoplanet by a mass measurement are costly or impossible for most targets. In the direct imaging case, the validation is not only to rule out false positives, but also to classify the planet type.

During the Exo-S mission's search phase, initial yet significant progress in planet identification will be based on the fundamental imaging measurements: 1) the apparent separation between the star and the source, 2) its relative brightness in the imaging bandpasses (color), and 3) its overall brightness relative to the star (reflectance).

Initial estimation of planet type will be done by colors in carefully chosen bands, although ambiguity will remain in some cases (see Figure 2.2-1). The interest is in separating

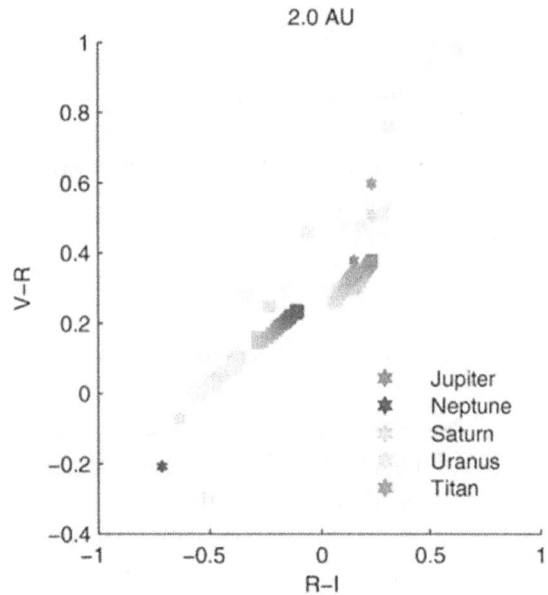

Figure 2.2-1. Color-color diagram from Cahoy et al. (2010) showing the locations of methane-rich solar system bodies (legend) and model planets (squares) placed at 2 AU. Red and magenta colors denote Jupiters with 1 and 3× enhancement over solar abundance in heavy elements. Blue and cyan are for 10× and 30× enhanced Neptunes. The color intensity fades as the model phase angles vary from 0° to 180° in 10° increments. Note that the anticipated variety of super Earth and sub-Neptune planet atmospheres is likely to clutter this diagram.

rocky worlds (of prime interest), sub-Neptunes (of high interest), and warm Jupiters from each other. Breaking the ambiguity, especially between sub-Neptunes and super Earths, is required to identify potential habitable zone rocky planets and likely requires follow-up spectroscopy.

In the second half of the Exo-S study, the STDT will research how far planet identification can go with color photometry, particularly with new classes of planets. For example, discerning between H_2-rich sub-Neptunes and N_2-CO_2-H_2O rocky planets may be possible from multi-color photometry (planets not shown in Figure 2.2-1). In addition, background objects and different planet classes have an expected range of colors and brightnesses that can overlap, leading to possible confusion (see Section 2.3).

A strategy for dealing with these uncertainties is to determine the probability that various types of sources can satisfy the observational measurements of apparent separation, color, and reflectance. In the second half of the Exo-S study, the STDT will generate probability density maps to formulate a more concrete observing strategy.

Direct imaging exoplanet science is a daunting task not afforded justice by a few outlined goals. Several pressing astrophysical questions have come to the forefront, including: how much can be learned about planets with limited spectral and temporal information; how planets can be efficiently distinguished from background sources; how stray light from binary stars should be handled; and how higher exozodi levels than the solar system's might impact the science harvest of a direct imaging mission. Answering these concerns will require a large-scale dedicated effort in the coming years.

2.2.2 Characterizing Exoplanets via Spectroscopy

2.2.2.1 Spectra Overview

In an ideal case, exoplanet atmospheric spectroscopy provides measurements for a variety of atmospheric characteristics, including atmospheric molecular composition and averaged vertical temperature profile. In turn, these measurements are expected to yield follow-on inferences related to the formation and evolution of planets and comparative planetology. To address these areas of study requires measurement of a number of planet atmosphere properties, including: the presence of water clouds and/or photochemical aerosols; the presence of water vapor; and determinations of planetary chemistry, including bulk elemental composition and redox state (e.g., CO_2-dominated or O2-dominated vs. H_2 dominated).

In reality, spectroscopy with a 1.1-m aperture telescope is limited in terms of programmatics: the number of candidate planets that can be spectroscopically followed up and the number of small planets for which

$R \sim 70$ spectroscopy can be obtained. These limitations come from long integration times within a limited mission lifetime, as well as planetary atmospheric chemistry: the same molecules (e.g., water vapor) are expected in planets of a variety of sizes and/or semi-major axes (see Figure 2.2-2).

Spectroscopy is of primary interest to help discriminate between planet types. Planets such as sub-Neptunes with thick H_2 atmospheres or envelopes have a larger atmospheric scale height than do rocky planets with thin atmospheres dominated by N_2 or CO_2. The larger scale height translates into deeper spectral features. Provided the flux from the bottom of the spectral features or a useful minimum flux can be measured, a sub-Neptune and a predominantly rocky super Earth can be distinguished.

Photo Removed Due to Copyright Restrictions

Figure 2.2-2. Differences and similarities in brightness and spectral features for a variety of exoplanet types. Optical reflectance spectra of a diverse suite of exoplanets is shown. The spectra have been convolved to a spectral resolution of 70. The Jupiter spectrum is based on the observed spectrum in Karkoschka (1994). The other two Jovian planet spectra are models from Cahoy et al. (2010). The Neptunian and water world spectra are models from Renyu Hu (personal communication). The Earth spectrum is a model developed to match Earth observations from the EPOXI mission (Robinson et at. 2011), while the super Earth is that model scaled by $(1.5\ R_E/1\ R_E)^2$. Image credit: A. Roberge.

Beyond identification of planet type, spectroscopy will enable identification of specific molecular features, including H_2O, CH_4, NH_3, O_2, and O_3.

The identification of these species will yield information on the chemical composition of the planet atmosphere. The goal for all planets is to take the chemical inventory, including absorbing gases and scattering aerosols, and estimate the bulk atmospheric elemental composition. Due to the constraints of mission lifetime and the required integration time for various planets, this goal will only be achieved for a subset of the newly discovered planets.

The details of what exactly can be inferred from direct-imaging spectra at $R \sim 70$ or lower has not yet received the same attention as the latest state-of-the-art spectral retrieval studies of transiting planets (e.g., Benneke and Seager 2012; Lee et al. 2013; and Line et al. 2014). The second part of the Exo-S STDT study will take a close look at what physical properties of the planet can actually be inferred from the expected spectral resolution and signal-to-noise ratio (SNR).

2.2.2.2 Exoplanet Main Atmospheric Spectral Features

A review of the main spectral features per planet type is useful to justify the wavelength range and spectral resolution capabilities of the Exo-S mission.

The Exo-S wavelength range for exoplanet atmospheric spectroscopy is 0.4–1.0 μm. This wavelength range encompasses absorption features from CH_4, NH_3, H_2O, O_2, and O_3, and permits detection of several CH_4 absorption features, required both for robust identification and for molecular abundance constraints. This is critical because the CH_4 is expected to be abundant on a variety of planets, ranging in size from sub-Neptune to Jupiter.

The long-wavelength cutoff can enable detection of the 0.94-μm H_2O band. This is the strongest water band at optical wavelengths in exoplanet atmospheres from giant planets to terrestrial planets. On planets with reducing atmospheres, as may be the case for sub-Neptunes, this water band is also the cleanest band, as the shorter wavelength H_2O bands are often combined with CH_4. The extension of the spectrum to 1.0 µm provides a chance to measure continuum on the long-wavelength side of the water band, which is optimal for quantification of water concentrations, based on the size and width of this feature.

The short wavelength cutoff provides for detection of Rayleigh scattering from gases, scattering from photochemical hazes, the effects of clouds, and possibly absorption characteristics of planetary surfaces. For example, the slope of the spectrum at short wavelengths may yield information on photochemical hazes such as H_2SO_4 (grey slope) and hydrocarbons (a redward slope).

The wavelength range will potentially allow discrimination between reducing atmospheres, similar to those seen on the ice giants in our solar system and expected for sub-Neptune planets, and oxygen- and water-dominated atmospheres like that presently found on Earth. In highly favorable cases, the spectral range also allows for a preliminary search for biosignature gases.

Key spectral features in planetary atmospheres, per planet type, are outlined. Many of these features are preserved at low spectral resolution as shown in Figures 2.2-3, 2.2-4, and 2.2-5.

Terrestrial Planet Spectra: Earth and Super Earths

Earth's reflected light spectrum is dominated by water vapor features, oxygen, and Rayleigh scattering. Earth's spectral features are shown in Figure 2.2-3

Other "theoretical" types of Earth and super Earth atmospheres will be studied in the second half of the STDT study (e.g., Figure 2.2-6). Relevant planet types include: an early Earth atmosphere, a hazy atmosphere from CH_4-rich atmospheres; a super Earth with a high concentration of CO_2; and an H_2-dominated super Earth with a biosphere,

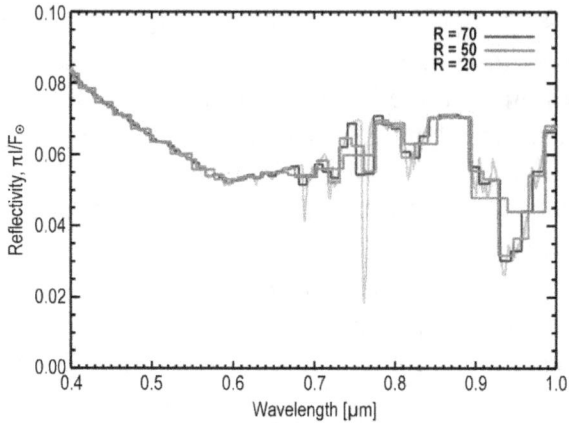

Figure 2.2-3. A theoretical Earth spectrum (grey) degraded to spectral resolutions of R = 20, 50, and 70. The oxygen A-band is seen at 0.76 µm, and the strongest water vapor band in this spectral wavelength range is seen at 0.94 µm.

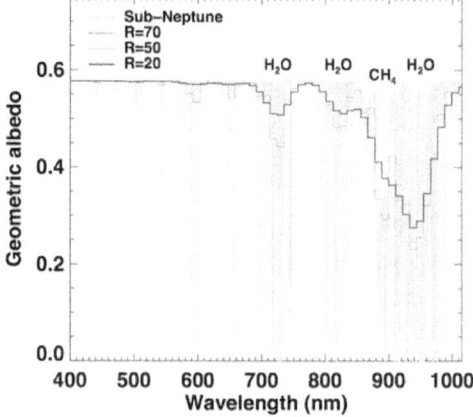

Figure 2.2-4. A theoretical spectrum for a metal-rich Neptune at 2 AU, valid also for smaller planets (i.e., sub-Neptunes) degraded to spectral resolutions of R = 20, 50, and 70. The strongest water vapor band in this spectral wavelength range is seen at 940 nm. CH_4 is needed to identify a planet as a sub-Neptune and not a rocky world with a thin atmosphere.

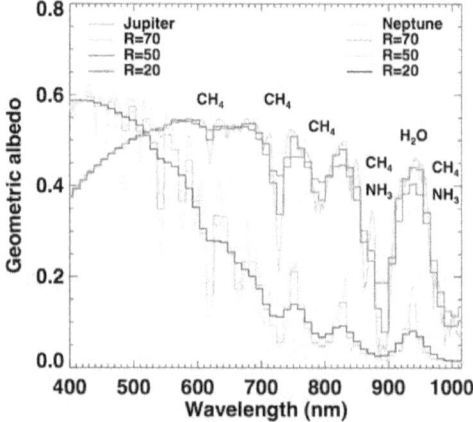

Figure 2.2-5. Jupiter and Neptune spectra (Karkoschska 1994) degraded to spectral resolutions of R = 20, 50, and 70. The strongest water vapor band in this spectral wavelength range is seen at 940 nm.

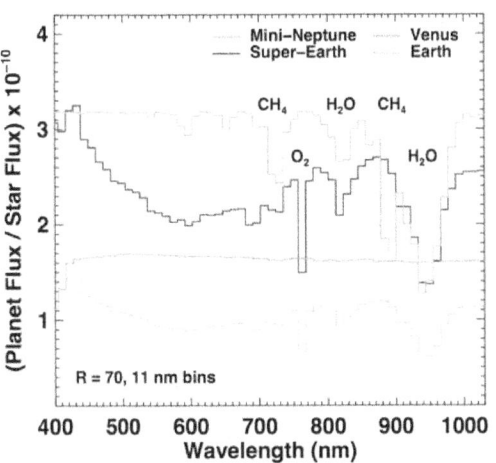

Figure 2.2-6. Simulated spectra of small planets. The Earth, Venus, and super Earth models are from the Virtual Planet Laboratory (VPL). The sub-Neptune model is from Renyu Hu (personal communication). The spectra have been convolved to $R = 70$ spectral resolution and re-binned onto a wavelength grid with 11 nanometer bins.

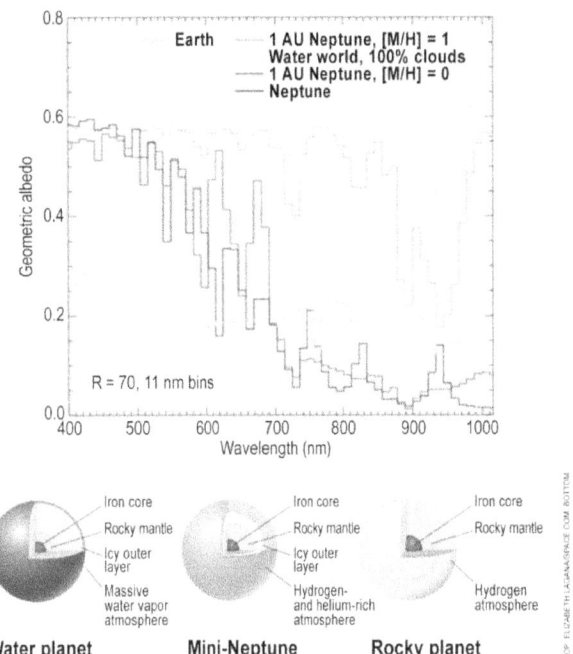

Figure 2.2-7. Geometric albedo spectra of modeled sub-Neptunes (R. Hu, private communication). Neptunes and sub-Neptunes will appear somewhat similar in terms of albedo spectra. The very natures of sub-Neptunes are unknown, and three possibilities are outlined in the cartoon diagram.

placed in an extended habitable zone around the star, far from the star but capable of maintaining liquid ocean water based on H_2 greenhouse gas warming.

Sub-Neptune Planet Spectra

The reflected light of spectra of sub-Neptune planets is anticipated to be dominated by methane, water vapor, and clouds, depending on the planet temperature and the planet atmosphere composition.

The sub-Neptune planets (planets ranging in size from about 1.75 to about 3 times the radius of Earth) are very low density, yet their actual composition is not known. The anticipated spectra of sub-Neptunes could ascertain the very nature of the low-density mysterious planets (Figure 2.2-7). Three extremes have been described: first, a water world planet composed of 50% water by mass with a thick steam atmosphere; second, a planet that is a smaller version of Neptune, i.e., an H/He atmosphere with high metallicity; and third, a rocky core surrounded by an H atmosphere, i.e., closer to solar metallicity and elemental abundances, with atmospheres similar to Neptune (Rogers and Seager 2010).

Giant Planet Spectra

The reflected light spectra of solar system giant planets (Figure 2.2-8) are dominated by strong methane absorption bands superimposed on bright continuum flux coming from clouds in the red and Rayleigh and/or haze scattering in the blue. Exoplanets slightly warmer than Jupiter may lack cloud decks, making atmospheric gases, such as water vapor, Na, and K, much more accessible to remote observations.

Most interesting is a comparison with the population of heretofore studied exoplanet spectra, that of the hot Jupiters. In addition, comparison with our own solar system giant planets and "warm" Jupiters near 1 AU will enable comparison of giant planets of very different kinds.

2.2.3 Known Giant Planets

Dozens of giant planets, that is planets with masses or sizes similar to Jupiter's, are already known to exist. Ground-based radial velocity surveys have uncovered most of them with a

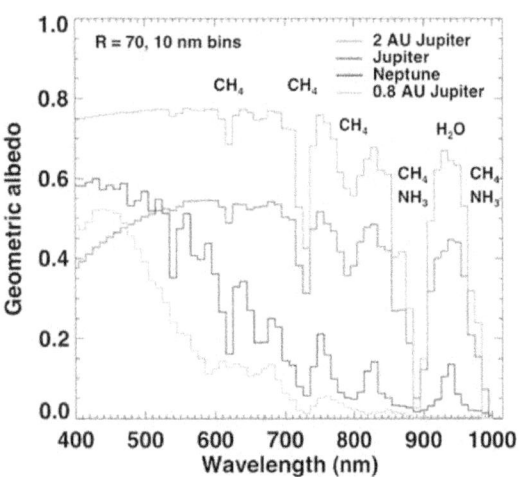

Figure 2.2-8. Giant planet spectra: geometric albedo spectra of real and modeled giant planets. The spectra have been convolved to $R = 70$ spectral resolution and re-binned onto a wavelength grid with 10 nanometer bins. The observed spectra of Jupiter and Neptune from Karkoscha (1999) are shown with red and blue lines, respectively. Two model giant planet spectra from Cahoy et al. (2010) are also plotted. They are warm Jupiter-like plants located 2 AU (orange line) and 0.8 AU (gray line) from a Sun-like star. The 2 AU Jupiter is very bright due to water clouds, while the 0.8 AU Jupiter is cloudless and darker.

few of interest discovered by ground-based direct imaging.

A few dozen giant planets whose minimum mass and orbits indicate they will be outside the Exo-S inner working angle (IWA) during the mission epoch are already known. A subset of these giant planets, 19 to be precise, have favorable locations in the sky and are accessible to the starshade as part of an efficient discovery-characterization program (see Section 3).

These known planets will be bright enough in reflected starlight that it will be possible to characterize them spectrally at $R \sim 70$. This represents an efficient, guaranteed science return for the Exo-S mission.

In addition to the spectroscopy-driven science outlined above, the prime motivation is to derive atmospheric molecular abundances and potentially metallicities to compare with the giant planets in our own solar system.

Masses of known RV planets will be determined by resolving the *sin(i)* orbital

inclination ambiguity. A first photometric measurement will pin down the orbital inclination, given the orbital elements already determined by the RV measurements.

2.2.4 Circumstellar Disk Science

The Exo-S mission preliminary DRM targets two known debris disk host stars (τ Ceti and ε Eridani), yielding images at unprecedented contrast levels. Approximately 6 additional targets are expected to yield images of debris disks. This estimate is based on a sensitive survey with the Herschel Space Telescope, which found that 20% of stars have cold dust disks $10\times$ as bright as the Kuiper Belt (Eiroa et al. 2013). While dust at 40 AU will be too faint to detect in most of these systems, a true Kuiper-Belt analog $\times 10$ would contain dust that the Exo-S mission could image in the <5 AU region, transported by comets and radiation drag. Moreover, in a true solar system analog, this <5 AU region contains dust produced by Trojan asteroids and small grains expelled from the main asteroid belt. New images of these dusty disks will yield a variety of scientific rewards.

2.2.4.1 Indirect Detection of Exoplanets

Debris dust coming from the destruction of asteroids and comets is a ubiquitous feature of planetary systems, including our own. Unfortunately, little is known about warm dust in the inner reaches of systems where habitable planets are expected to reside. In the solar system, the warm dust interior to the asteroid belt is called zodiacal dust, and it appears to come largely from the evaporation of comets (Nesvorny et al. 2010).

Debris disk morphologies reflect the gravitational perturbations of the planets they contain (e.g., Dawson et al. 2011); observing structures in debris disks offers a potentially powerful indirect tool for finding planets and constraining their masses and orbital parameters. For example, Neptunes and super Earths orbiting at semi-major axes beyond roughly 15 AU have orbits too long to permit

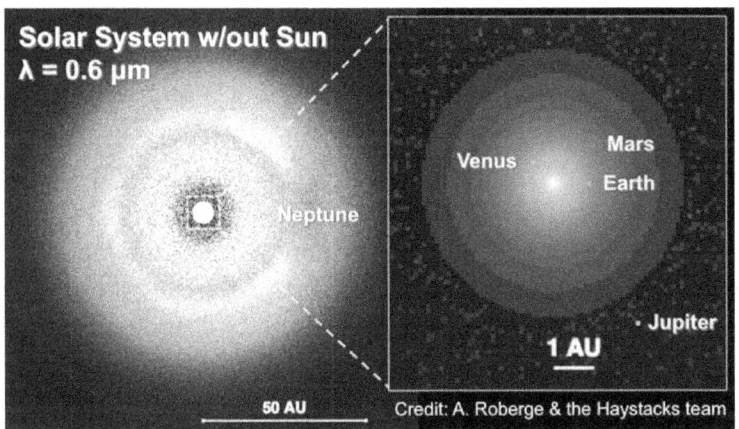

Figure 2.2-9. A slice from a high-fidelity model of the solar system dust complex (Kuchner and Stark 2010). For ease of viewing, the Sun is not included (as if it had been perfectly suppressed with a high-contrast instrument). The image on the left shows the entire solar system, the image on the right shows the inner 5 AU, with a zodiacal cloud model from ZODIPIC, based on Kelsall et al. (1988). At both distance scales, and at all wavelengths, the most conspicuous feature is the haze of emission coming from interplanetary dust. The partial ring in the outer solar system dust is caused by the dynamical influence of Neptune. Different planetary configurations and masses would produce a different diagnostic dust distribution.

detection via RV, transit, or astrometric techniques within a human lifetime. They are also too faint (~10^{-12} in planet-star flux contrast) to detect directly with any known technique. The only way to detect true Neptune analogs around nearby stars may be to study the structures of debris disks.

Observations of τ Ceti and ε Eridani with the Exo-S probe will be well suited to this task. Imaging of ε Eridani with Spitzer suggests a complex system of possible five dust components, including belts at 3 AU and 20 AU, which the Exo-S mission will easily resolve (Backman et al. 2009). The τ Ceti system apparently sports both a hot dust component (di Folco et al. 2007) and a cold dust component (Greaves et al. 2004), both of which are also accessible to the Exo-S mission. This complex structure hints that there may be planets sculpting the belts, whose properties we can constrain using images from Exo-S.

Models suggest that in massive debris disks like those around τ Ceti and ε Eridani, Neptune-mass planets can sculpt the disk into eccentric rings. Observing these rings can constrain the planet's mass and orbital eccentricity (e.g., Chiang et al. 2009).

Observations of new habitable zone dust clouds by Exo-S will provide further opportunities to harness planet-disk interactions to aid with planet detection. Models suggest that a two Earth-mass planet can produce a detectable resonant structure in an exozodiacal cloud similar to the solar zodiacal cloud at 1 AU, whose morphology can constrain the planet's mass and eccentricity (Figure 2.2-9; Stark and Kuchner 2008). The signatures get stronger further from the star and in dust clouds dominated by larger grains, or generated by dynamically cooler sources. So under some circumstances, a Mars-mass planet may even yield a detectable signature, a ring with a gap at the planet's current location.

2.2.4.2 Transport-Dominated Disks

The disks imaged by the Exo-S mission, which will include fainter disks than any previously imaged, should be the first to trace a dramatic transition in disk physics. More massive disks—all the ones currently known—are *collision dominated*; the dust grains we observe are mainly destroyed by collisions with other grains. But disks with optical depth less than v_K/c, where v_K is the local Keplerian speed, are predicted to be *transport dominated*, meaning that grain-grain collisions are rare enough that grains can flow throughout the planetary system under that influence of radiation drag forces before they are sublimated in the star's corona or ejected from the system by an encounter with a planet. This transition between collision dominated and transport dominated corresponds to a contrast level of ~10^{-7} in the habitable zone.

What's important about observing this predicted transition is that the physics of transport-dominated disks is much simpler than that of collision-dominated disks, so it is easier to interpret the morphology of these disks in terms of the properties of hidden planets that are perturbing them. Modeling the dust distribution in collision-dominated disks requires understanding the details of collisional processing and the distribution of planetesimals, remnants of the complex process of planet formation and migration (e.g., Nesvold et al. 2013). But dust transported away from its source can be modeled with a simple n-body integrator, and the range of possible planet-dust interactions is already understood (e.g., Kuchner and Holman 2003). Presently, the only known example of a transport-dominated debris disk is the solar system dust complex.

2.3   Astrophysical Contaminants

Deep imaging in close proximity to nearby stars will reveal not only planetary companions, but a plethora of background sources and exozodiacal light with unresolved structure. How can these astrophysical contaminants be efficiently distinguished from planets? In this section, background sources are reviewed and five mitigation strategies are listed that may be appropriate for a probe-class mission. The handling of astrophysical contaminants will be explored in greater depth during the second half of the STDT-S study, and beyond.

2.3.1   *Exozodiacal Clumps and Background Objects*

The exozodiacal dust levels around nearby stars will be as important to the success of efforts to characterize Earth-like exoplanets as the fraction of stars with potentially habitable planets ($\eta_\oplus$). Exozodiacal dust complicates direct imaging of exoplanets in two ways: (1) as a source of photon noise and (2) as a source of confusion due to unresolved structures that could masquerade as planets (Roberge et al. 2012). Background flux from similar exozodiacal dust (exozodi) in other systems

will likely dominate the signal of an Earth-analog exoplanet in direct images and spectra, even if exozodi levels are no greater than the solar system level.

Currently, little is known about the dust surrounding most Exo-S targets. This situation will be improved within the next few years by a new ground-based survey for exozodi around nearby stars (called "HOSTS") using the Large Binocular Telescope Interferometer (LBTI) (Hinz 2013). The HOSTS survey will measure the integrated 10-µm thermal emission from warm dust down to about 10 times the solar system zodiacal dust level at 1 AU scales.

The LBTI HOSTS survey, however, will not address two additional aspects of the exozodiacal dust problem for a future exo-Earth imager. First is the issue of how to convert the observed 10-µm dust emission to an optical surface brightness: a value for the dust albedo must be adopted in order to predict the exozodi background that Exo-S will encounter. Secondly, the HOSTS survey data, integrated over the fringe pattern of a nulling interferometer, will provide little information on the spatial distribution of the exozodiacal dust. While the HOSTS survey will certainly aid mission planning for Exo-S in helping to constrain overall dust levels, the Exo-S probe itself will be sensitive to disks as faint as 0.1 times as bright as the solar zodiacal cloud. As an exceptionally powerful probe of this astrophysical noise source, Exo-S will provide information that will help guide planet-imaging missions for decades to come.

Due to the unknown and likely presence of dust and unresolved dust structures surrounding Exo-S targets, the current Exo-S DRM provides for follow-up spectroscopy of all exo-Earth candidate sources during the second visit. The Exo-S STDT will investigate how the mission science return declines with increasing dust levels, with the aim to devise strategies to disentangle dust structures and planet candidates.

2.3.2 Scattered Light from Companion Stars

Many stars of highest interest have known stellar companions, and their presence will inevitably introduce scattered light into the image plane. Assuming that this represents a smooth background, a longer exposure time will be required in order to reach the desired signal to noise for detecting Earth or Jupiter analogs. This will *not* rule out all binary stars, but targets with bright companion stars at small separations may be severely compromised.

In work done prior to this study for a 4-m class mission, the relative increase in exposure time to detect an Earth-like planet in the habitable zone was calculated for stars that have separation and magnitude data for companions in the Washington Double Star catalog (priv. comm., M. Turnbull and C. Noecker). These sample calculations are shown in Figure 2.3-1 to illustrate how stellar multiplicity can affect target selection for a direct imaging mission. The calculations were carried out using Spyak and Wolfe (1992) and Kuhn and Hawley (1999) stray light model predictions, and here, only the more severe Spyak and Wolfe (1992) prediction is plotted.

The key qualitative findings from these calculations are:

1. According to these model predictions, the presence of a companion has little or no effect on integration time for angular separation more than ~40 arcsec. Systems such as α Cen A and B are potentially quite problematic due to the combination of small angular separation and similar brightness of the two components.

2. At smaller separations, stray light from companion stars must be modeled in detail. Many binary star systems with separations less than 40 arcsec will still remain viable targets (e.g., η Cas A) if they are very nearby and/or the magnitude difference of the two components is large.

During the remainder of the Exo-S study, and as a general course of preparatory science, stray light calculations will be carried out to assess how exposure times will be affected for all candidate targets that have stellar companions. These calculations will consider the ever-changing separations of the two components, the effects of additional stars in systems with three or more components, and

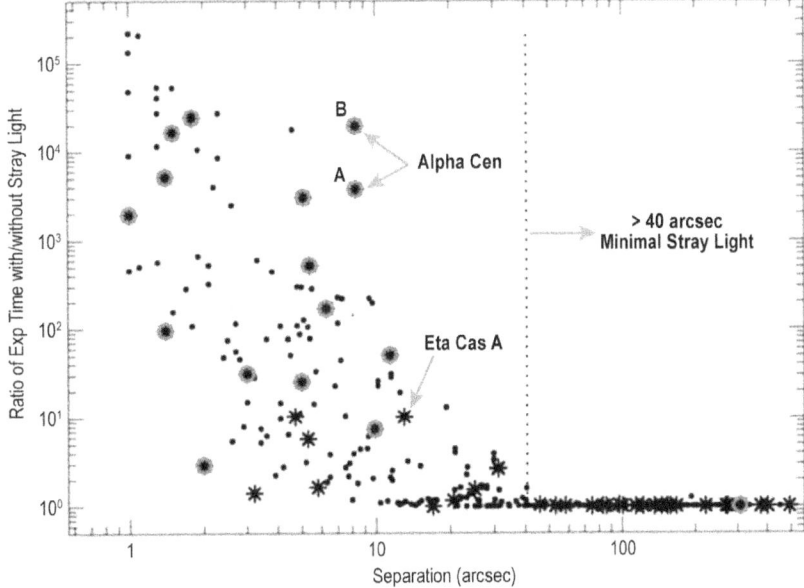

Figure 2.3-1. The expected relative increase in exposure times to detect exo-Earths for stars with bright companions. Asterisks indicate stars that have the highest habitable zone completeness. Red-encircled targets are those whose exposure times may be prohibitive when considering stray light.

whether both components of very tight (spectroscopic) binaries can be effectively nulled simultaneously.

The current Exo-S DRM avoids targets where stellar companions are likely to introduce high levels of stray light.

2.3.3 Background Stars

At low galactic latitudes, the appearance of background stars in the Exo-S planet search field is inevitable. As the starshade acquires its target, previously unseen "bright" background stars (V = 10–20 mag) will emerge within the Exo-S field of view (FOV) and potentially overwhelm portions of the detector for planetary companions. Meanwhile, fainter background stars may masquerade as planet candidates. Star counts are expected to be near 100,000 stars per square degree for 10 < V < 20 mag (or about 25 bright stars in every planet-finding field), and 500,000 stars per square degree for the range 20 < V < 25 (i.e., as many as 100 faint stars in each planet-finding field; Binney and Merrifield 1998). At V > 25, the star counts must eventually decline due to the finite size of the galaxy, but variability in galactic extinction make this somewhat unpredictable for any given field of view.

Figure 2.3-2 shows a 5'×5' field of background stars near η Cas A (HIP 3821) to a limiting magnitude near V ~ 22 (from the Space Telescope Science Institute [STScI] Digitized Sky Survey, POSS2/UKST [Second Palomar Sky Survey/UK Schmidt Telescope] blue image). This target is currently part of the Exo-S DRM for the exo-Earth search. To mitigate the damage done by brighter background stars, it may be necessary to co-add short exposures, or to commence observations with relatively short exposures and increase exposure times if no "bright" background stars are present. Another approach is simply to design a mission program that avoids targets in regions of known or suspected high stellar background density. Finally, many Exo-S targets are high

Photo Removed Due to Copyright Restrictions

Figure 2.3-2. A 5' × 5' field of background stars near η Cas A (HIP 3821) to a limiting magnitude near V ~ 22 (from the STScI Digitized Sky Survey, POSS blue image), illustrating the number of "bright" background stars likely to appear in planet search images. The Exo-S FOV is 1'× 1'. This target star is currently part of the Exo-S DRM for the exo-Earth search.

proper motion stars, and mission programming would be aided by deep imaging with the Hubble Space Telescope (HST) or James Webb Space Telescope (JWST), several years in advance of the direct imaging mission. In this case, the exact distribution of background sources could be characterized and the observing protocol adapted accordingly.

Fainter background stars (V > 25) will be significantly reddened, and therefore it may be possible upon the first visit to distinguish them from planet candidates through broadband color data. However, it is clear that in broadband measurements, planets shining in reflected starlight will display a wide range of colors, just as stars do. The Exo-S Final Report will examine the multi-color-magnitude phase space of planets and background stars in order to assess how efficiently planet candidates can be identified in the first visit.

2.3.4 Extragalactic Sources

Away from the galactic plane, the Exo-S probe will essentially see a Hubble Ultra Deep Field

(HUDF) for every imaged target star and planet, and the primary concern is a veritable ocean of (1) brighter (V < 28) extended galaxies and (2) ultra-faint unresolved galaxies (mostly V > 28) displaying a wide range of colors (-2 < B–V < 2) and tending toward the extreme blue at the faintest magnitudes (Coe et al. 2006; Pirzkal et al. 2005). In cases where an unresolved background galaxy falls within the expected multi-color range for planets, proper motion discrepancies and spectroscopic follow-up will be able to disambiguate these sources upon the second visit.

Figure 2.3-3 shows a one arcmin^2 FOV at high Galactic latitude, extracted from the Hubble Extreme Deep Field (XDF, Illingworth et al. 2013). The deepest part of the XDF has a limiting magnitude near V ~ 31, and contains 7,121 galaxies above the 5-sigma significance level in ~4.7 arcmin^2. This corresponds to ~1,500 galaxies in the 1 arcmin^2 Exo-S FOV. Galaxy counts from other surveys (Windhorst et al. 2011) indicate that we should expect a few dozen galaxies per FOV at V < 25, and these brighter, extended, non-uniform sources could make planet detection difficult wherever they dominate the signal. Meanwhile, examination of the XDF reveals significant image crowding at V ~ 30, where 45% of the pixels contain galaxy light (Koekemoer et al. 2013).

The large number of background sources, including stars within our own Galaxy and galaxies beyond, necessitate a second visit for every exo-Earth search target. This is currently included in the Exo-S DRM. Deep field observations using HST or JWST to observe along the future path of high proper motion targets may be useful in removing non-common proper motion background sources from planet search images.

2.3.5 Summary of Observing Protocols

To summarize, astrophysical sources such as exozodiacal light and bright off-axis stellar companions serve to increase the required exposure times to detect exoplanets.

Photo Removed Due to Copyright Restrictions

Figure 2.3-3. A one arcmin^2 FOV (corresponding to the planet detection field for Exo-S), extracted from XDF (Illingworth et al. 2013). About 1,500 extragalactic sources down to V ~ 31 are present in this image.

Meanwhile, exozodiacal dust clumps, background stars, and faint extragalactic sources may pose as planets and must be ruled out through proper motion discrepancies, broadband color, or spectroscopic analysis.

The observing protocol for Exo-S incorporates all of these elements for exo-Earth search targets. A preliminary process for Exo-S observation is:

1. Preparatory observations: in the very near term, use HST observations to observe along the future path of high priority, high proper motion targets. Many Exo-S targets move 1 arcsec per year or more, and one HST orbit of observing time can detect sources along this path down to V ~ 26. Targets likely to have prohibitively complicated background may be removed from the program in favor of targets with cleaner fields.

2. Once the Exo-S mission has commenced, on visit #1, obtain images in 3–4 broadband colors for a cursory look at all faint sources in the field. Sources very different in color from the target star are

likely to be background objects. Sources that did not appear in prior imaging efforts may be of particular interest for follow-up spectroscopy.

3. On visit #2 (approximately 12 months later), obtain broadband colors again and compare fields for proper motion companions. Broadband color should aid in verifying that sources are the same companions observed previously.

4. On visit #2, commence with spectroscopy for sources that are proper motion companions to the target star. Spectroscopy will confirm that planets are shining in reflected starlight as well as reveal signatures of key atmospheric species in the planets themselves.

2.3.6 Summary of Preparatory Science Recommendations

For the remaining term of the Exo-S STDT study and beyond, high priority tasks for efficiently handling astrophysical contaminants include:

1. Calculation of exposure times including the effect of stray light from binary companions

2. Identification of multi-color bands that can efficiently rule out background sources and shed light upon planet atmospheres

3. Obtaining deep background fields (e.g., with HST) along the future path of high priority, high proper motion targets

3   Design Reference Mission

A probe-class mission is defined with high value science using a starshade for direct exoplanet imaging.

This section presents a preliminary Design Reference Mission (DRM) for the baseline Exo-S mission. In the first 22 months of a 3-year prime mission, 3 tiers of target stars will be observed:

1. Tier 1: 20 stars with high probability of detecting Earth twins present in the habitable zone, or brighter planets such as sub-Neptunes.
2. Tier 2: 17 stars hosting 19 known radial velocity (RV) planets, plus potential new planets.
3. Tier 3: 16 stars with high probability of detecting Jupiter twins.

Achieving this high value science with a small and conventional telescope demonstrates two inherent starshade strengths. First, the inner working angle (IWA) is solely a function of starshade size and telescope separation distance (see Figure 1.2-1) and this enables use of a *small* telescope. Second, starlight is suppressed prior to entering the telescope and this enables use of a *conventional* telescope with a simple high-throughput instrument. There is no wavefront correction system and the outer working angle (OWA) is unlimited, providing access to large portions of each planetary system.

This combination of deep starlight suppression and high throughput yields reasonable integration times. Tier 1 Earth twin candidates are observed in 3 colors with an average integration time of 3.4 days. Of the 19 Tier 2 planets, 12 are characterized to $R = 70$ with an average integration time of just 2 days, while 7 are characterized to $R = 50$ with an average integration time of < 6 days. The principle limitation of this small telescope option is the spectral resolution achievable on Earth twins with a reasonable allocation of mission times, as detailed in this section.

This DRM also demonstrates the mitigation of an inherent starshade weakness, namely the limited observation time as a result of the necessary large translational retargeting maneuvers. The use of high efficiency ion propulsion, with commercially available equipment, reduces the required propellant mass to practical levels (<70 kg). The fractional observing time is 25%, which is in line with an experiment sharing a multi-purpose telescope.

The following sections detail key system trades, specific target lists, the DRM approach, and, finally, the baseline observation sequence.

3.1   Key System Performance Parameters

An iterative mission-level trade study led to a mission with an excellent balance between observational performance and hardware capabilities. The first point to consider is that the starshade diameter is limited to not exceed 34 m, for reasons of manufacturability. The most critical observational performance parameters are IWA and photometric sensitivity.

The IWA drives obscurational search completeness. Figure 3.1-1 shows the influence of IWA on the number of targets available, with a representative photometric sensitivity, limΔmag, of 25 magnitudes. limΔmag is the planet contrast at the threshold of detectability. The blue points represent all

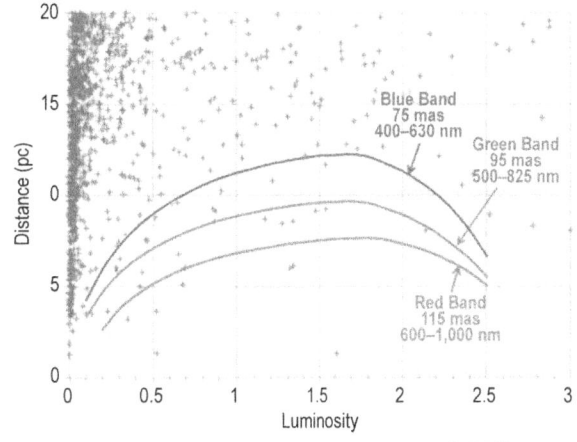

Figure 3.1-1. Targets vs. IWA and bandpass at 25% Earth twin search completeness and limΔmag = 25. Luminosity is in units of solar luminosity.

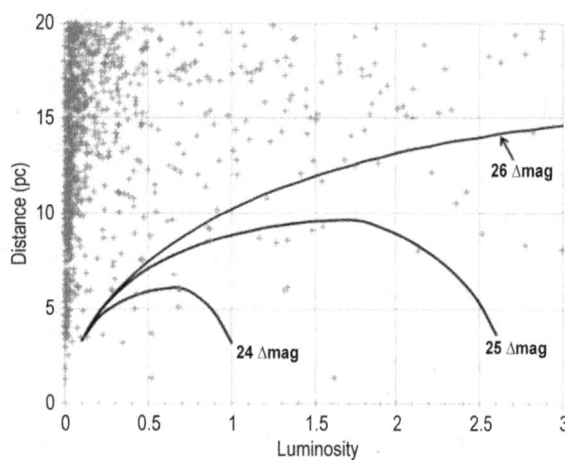

Figure 3.1-2. Targets vs. lim∆mag at 25% Earth twin search completeness and 95 mas IWA. Luminosity is in units of solar luminosity.

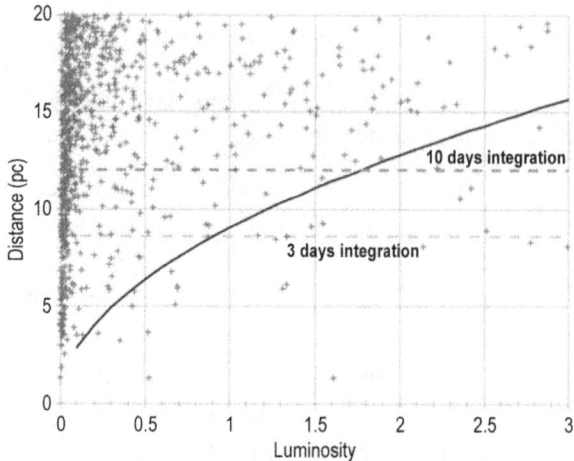

Figure 3.1-3. Earth twin targets at 25% search completeness with 95 mas IWA and ∆mag = 25.1 + 2.5 logL. The integration timelines show the maximum distance to characterize an Earth twin to SNR = 5 in a 510–825 nm band in the given number of days. Luminosity is in units of solar luminosity.

stars within 20 pc of the Sun having luminosity < 3, while the contours represent constant Earth twin search completeness at 25%. This is the probability of detecting an Earth twin if present in the habitable zone (here assumed to be 0.7–1.4 AU scaled by the square root of stellar luminosity). Stars below the curves have at least 25% observational completeness. Three observing bands (blue, green, and red) are defined, each one matching the starshade bandpass for a telescope separation distance inversely proportional to wavelength. Each band preserves a constant number of Fresnel zones across the occulter. Green is the primary band and it captures a number of spectral features of interest (e.g., O_3, O_2, H_2O, and CH_4). The red band captures additional spectral features at longer wavelengths, including water and CO_2, but at the expense of a larger IWA. The blue band offers the best habitable zone access with the smallest IWA.

Figure 3.1-2 shows the influence of lim∆mag on the number of accessible target with an IWA of 95 mas. Again, the contours represent constant Earth twin search completeness at 25%. However, with calibratable systematic noise, lim∆mag becomes photometrically limited and therefore adjustable with integration time. One approach is to adjust photometric sensitivity as: ∆mag = 25.1 + 2.5 log L, where L is stellar luminosity. Figure 3.1-3 shows this curve and integration times of 3 and 10 days. Again, the contour represents 25% Earth twin search completeness for a 95 mas IWA.

For Tier 3 targets (Jupiter-twin candidates), the limiting sensitivity is set at lim∆mag = 22.5. Figure 3.1-4 shows a contour of constant Jupiter-twin search completeness at 25% for an IWA of 95 mas. More than 100 candidate Tier 3 targets are available. Tier 2 known RV planets are also shown in Figure 3.1-4 (see red diamonds). Assuming Jupiter size and albedo, the limiting sensitivity

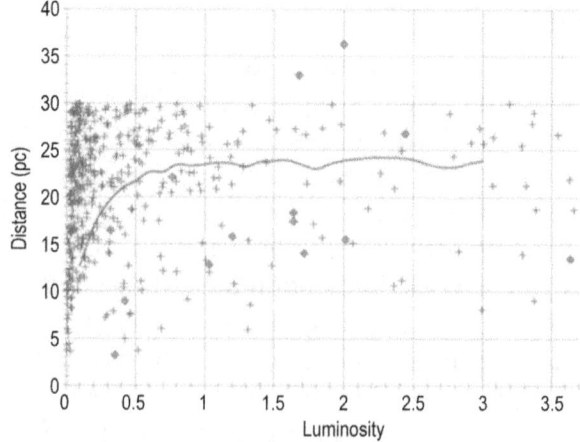

Figure 3.1-4. Jupiter-twin targets at 25% search completeness with 95 mas IWA and ∆mag = 22.5. Luminosity is in units of solar luminosity.

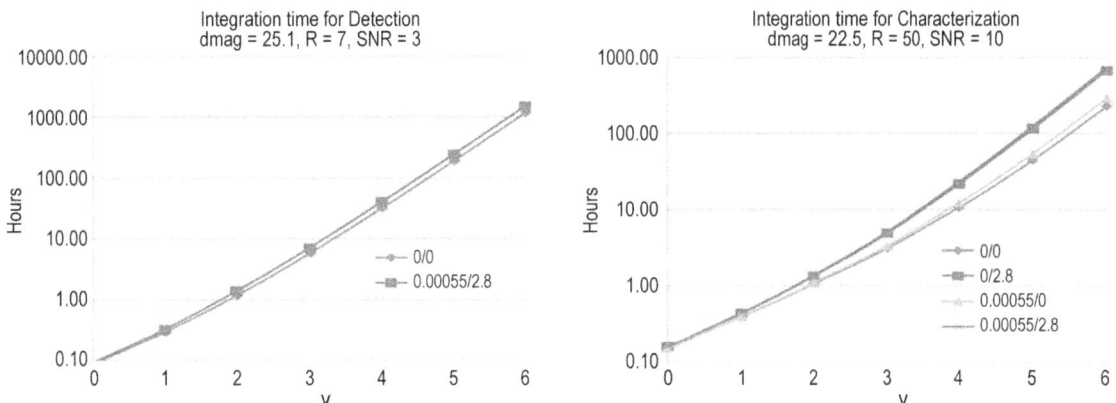

Figure 3.2-1. Integration time for detection of Earth twins (left) and characterization of Jupiters (right). Curves are labeled with dark current (e-/pix/sec) and read noise (e-). The assumed sharpness criterion for detection is 0.04, while it is 0.2 for spectral characterization. There are 2,000 s between readouts.

is set to match each target based on the position derived from its RV orbit.

3.2   Integration Times

The baseline imaging and spectroscopy detectors are e2V CCD-273 charge coupled devices (CCDs) developed for the Euclid mission. They have a target read noise of 2.8 e-/pixel and specified dark current of 0.00055 e-/pixel/s at 150K (Endicott et al. 2012). Figure 3.2-1 shows the impact on integration time relative to a noise-free detector. The CCD noise is seen most strongly during spectral characterization where the natural background per channel is reduced by the spectral resolution. While a zero-noise detector is desirable, it is not essential, and for this DRM, the noise penalty of the commercial device is acceptable.

Our integration time calculations include the local zodiacal light (23 mag/sq. arcsec) and 3 zodis of exozodiacal light (3 × 22.5 mag/sq. arcsec), for a combined surface brightness of 20.9 mag/sq. arcsec. For Jupiters, the exozodi density is conservatively assumed to remain the same even at the few-AU orbits of the planets.

Detection observations are made simultaneously in 3 bands each with spectral resolution between 6–9 depending on the final choice of filters. Integration times are calculated using average spectral resolution of $R = 7$ per band. Integrations continue until the

Table 3.2-1. System parameters.

Input Parameter	Units	Earths	Jupiters
Limiting delta magnitude		25.1	22.5
Spectral resolution		7	50–70
Telescope throughput		0.8	0.8
Instrument throughput		0.5	0.4
SNR		3	10
CCD dark rate	ph/s	0.00055	0.00055
CCD read noise	e/read	2.8	2.8
Single-frame exposure time	sec	2000	2000
Design contrast (residual starlight)		1.00E-10	1.00E-10
Telescope diameter	m	1.1	1.1
Quantum efficiency (QE)		0.8	0.8
Total zodiacal surface brightness	mag/sq. arcsec	20.9	20.9
Detection wavelength	m	6.00E-07	6.00E-0.7
Sharpness		0.08	0.28

photometric signal-to-noise ratio (SNR) of the planet to the background noise is 3 per band. The combined broad-band image then has photometric SNR of ~3*sqrt(3) = 5.2.

Characterization observations require higher SNR because spectral features may be weak. SNR = 10 is used for the continuum in characterization observations. For the bright Jupiters, the instrument spectral resolution is $R = 70$, while for fainter ones, $R = 50$ to limit integration times to <10 days.

Other system parameters are summarized in Table 3.2-1. The telescope has just two reflections and the instrument has no more than 5 reflections per path. The optical throughput is conservatively estimated to be

40%. The detector quantum efficiency (QE) is assumed to be 0.8 based on CCD-273 (Endicott et al. 2012). Finally, the "sharpness" criterion for imaging is based on Nyquist sampling of a diffraction limited Airy pattern, leading to sharpness = 0.08 and an effective number of pixels N_{pix} = 1/sharpness = 13 (Brown et al. 2006). For spectral characterization, the signal is integrated in the direction orthogonal to the dispersion, sampling with N_{pix} = 4 pixels per spectral resolution element and an effective 1-dimensional sharpness of sqrt(0.08) = 0.28.

3.3 Target Lists

Our target list excludes binary stars with close companions. Alpha Centauri A/B are specifically excluded because their large size results in the shadow converging in front of the telescope. This applies at the standard distances set for the three observing bands. It may be possible to carry out the Alpha Cen observations with the starshade moved closer to the telescope and this will be studied further and addressed in the Final Report.

3.3.1 Candidate Earth Twins

The Exo-S DRM surveys 20 targets that offer a high probability of detecting an Earth twin. Table 3.3-1 summarizes the targets. Additional targets are available to observe later in the mission but they were not scheduled to be observed in the first 22 months due to Sun-pointing constraints complicated by other targets at similar ecliptic longitude.

The average search completeness is 38% and average integration time for detection is 3.4 days. For the Earth twin search, limΔmag is fixed at 25.1 rather than scaling it with luminosity. This keeps integration times shorter for high-luminosity stars at the expense of completeness. This approach will be revisited in the Final Report to find optimal observing times that maximize the overall program completeness as has been studied with the Terrestrial Planet Finder Coronagraph DRM (Brown 2005). Note that, assuming limΔmag = 25.1, it will also be possible to detect sub-Neptunes (~2.5 Earth radii) out to about 3 AU around these targets. Meanwhile, Neptune- (5 Earth radii) and Jupiter-sized

Table 3.3-1. Target list for Earth twin survey.

Target #	Name	HIP	Spectral Type	[Fe/H]	V (mag)	Distance (pc)	B-V Color Index	Search Completeness	Detection Time (days)	R for 30-Day Integration
1	τ Ceti	8102	G8.5V	-0.52	3.49	3.65	0.73	0.72	0.70	40
2	82 Eridani	15510	G8.0V	-0.41	4.26	6.04	0.71	0.65	2.74	17
3	σ Draconis	96100	G9.0V	-0.19	4.67	5.75	0.79	0.62	5.71	10
4	η Cassiopei A	3821	G3V	-0.25	3.45	5.94	0.57	0.37	0.66	41
5	GL 189	23693	F6/7V	____	4.71	11.65	0.53	0.32	6.16	9
6	GL 150	17378	K0IV	0.16	3.52	9.04	0.93	0.14	0.74	39
7	GL 107A	12777	F7V	0.06	4.10	11.13	0.49	0.24	2.05	20
8	GL 124	14632	G0V	0.16	4.05	10.54	0.60	0.26	1.88	21
9	o 2 Eridani	19849	K0.5V	-0.28	4.43	4.98	0.82	0.54	3.72	14
10	GL 216A	27072	F7V	____	3.59	8.93	0.48	0.26	0.83	36
11	β CVn	61317	G0V	-0.16	4.24	8.44	0.59	0.36	2.65	17
12	GL 502	64394	G0V	0.07	4.24	9.13	0.57	0.32	2.63	17
13	GL 442A	57443	G3/5V	-0.33	4.89	9.22	0.66	0.41	8.53	7
14	61 Vir	64924	G5V	0.05	4.74	8.56	0.71	0.48	6.50	9
15	GL 598	77257	G0Vvar	0.05	4.41	12.12	0.60	0.24	3.61	14
16	GL 695A	86974	G5IV	____	3.41	8.31	0.75	0.21	0.60	44
17	GL 231	29271	G6V	0.09	5.08	10.20	0.71	0.33	12.05	6
18	δ Pavonis	99240	G8.0IV	0.33	3.53	6.11	0.76	0.46	0.75	38
19	GL 827	105858	F7V	____	4.22	9.26	0.47	0.31	2.55	17
20	GL 17	1599	G0V	-0.22	4.23	8.59	0.58	0.35	2.60	17

Table 3.3-2. Known RV planets spectrally characterized with this DRM.

Target #	Name	HIP	Spectral Type	[Fe/H]	V (mag)	Distance (pc)	B-V Color Index	ΔM_v	R	Charact. Time (days)
21	HD 11964 b	9094	G5V	___	6.42	33.00	0.84	21.17	50	6.0
22	υ And d	7513	F8V	0.15	4.10	13.49	0.54	22.45	70	3.3
23	e Eridani b	16537	K2V	-0.03	3.71	3.21	0.88	21.55	70	1.2
24	γ Cep b	116727	K1III	___	3.21	14.10	1.03	20.39	70	<1
25	7 CMa b	31592	K1III	___	3.95	19.75	1.06	20.34	70	<1
26	Pollux b	37826	K0III	___	1.15	10.36	0.99	20.86	70	<1
27	47 Uma b, c	53721	G0V	0.04	5.03	14.06	0.61	20.24	70	<1
28	HD 128311 c	71395	K3-V	0.2	7.49	16.50	0.97	20.21	50	7.2
29	HD 147513 b	80337	G1V	0.09	5.37	12.78	0.63	19.08	70	<1
30	μ Ara b, c	86796	G3IV-V	0.29	5.12	15.51	0.69	22.20	50	3.8
31	HD 164922 b	88348	G9V	0.17	7.01	22.12	0.80	20.29	50	3.6
32	HD 39091 b	26394	G0V	0.05	5.65	18.32	0.60	21.46	70	4.6
33	HD 169830 c	90485	K1III	___	5.90	36.32	0.48	21.51	50	4.4
34	HD 192310 c	99825	K2+V	0.02	5.72	8.91	0.91	19.00	70	<1
35	HD 216437 b	113137	G1V	0.22	6.04	26.75	0.66	20.77	70	5.0
36	HD 190360 b	98767	G7IV-V	0.21	5.73	15.86	0.75	22.06	50	8.6
37	HD 10647 b	7978	F9V	-0.08	5.52	17.43	0.53	20.37	70	3.4

(10 Earth radii) planets will be detectable out to ~5 AU and 12 AU, respectively.

Table 3.3-1 also shows the spectral resolution achievable for a representative observation time of 30 days. Note that if characterizing a particular confirmed Earth twin is a high priority after the discovery part of the mission, then longer observation times are certainly possible. The limit varies with ecliptic latitude of the target, but 60 days is reasonable if not limited by the planet's orbital motion.

Three of the Earth twin candidate targets have known Jupiter-sized planets present inside of 0.5 AU. These are: 83 Eridani (3 planets), G1 442A (1 planet), and 61 Virginis (3 planets). It is an open question whether these systems with close-in giants can have Earths in the habitable zone because it's possible the giants migrated into those radii, sweeping up everything in the habitable zone in the process. But if terrestrial planets could form afterwards, Exo-S would be capable of detecting them.

3.3.2 Known RV Planets

The Exo-S DRM spectrally characterizes 19 known RV planets orbiting 17 stars to a spectral resolution ($\lambda/\Delta\lambda$) of 70, or in a few cases, 50 (Table 3.3-2). A total of 61 RV planets are accessible with an IWA of 95 mas. R. Brown (personal communication) has used the RV orbital data to determine the subset of planets that will be accessible in the 2024–2026 timeframe. The list is further restricted to those targets with characterization times less than 10 days. Integration times are set to achieve the limΔmag that matches each planet's predicted brightness assuming the planet matches Jupiter in size and albedo. The average integration time is 3.4 days. In the Final Report, the allocation of mission time will be revisited and this may lead to additional known RV planets.

It is important to note that a single observation of the system measures the inclination and determines the planet mass. Exo-S will measure the spectra and masses of no less than 19 exoplanets.

3.3.3 Jupiter Twin Survey

There are many stars around which Exo-S can quickly discover and characterize Jupiter-sized planets. These form the third tier of targets and they are selected based on their location and observational completeness. Of 56 stars determined to have completeness for Jupiter detection > 35%, detection times < 1 day, and characterization times at $R = 30$ of less than 12

Table 3.3-3. Target list for Jupiter candidates.

Target #	Name	HIP	Spectral Type	[Fe/H]	V (mag)	Distance (pc)	B-V Color Index	Search Completeness	Detection Time (days)	R for 30-day Integration
38	GL 542.1A	69965	F7(W)F3V	___	5.88	18.03	0.48	0.369	0.44	55
39	GL 582	75181	G3/5V	-0.34	5.65	14.81	0.64	0.393	0.29	70
40	GL 705.1	89042	G1V	___	5.47	17.61	0.59	0.407	0.21	84
41	GL 55	5862	G0V	0.16	4.96	15.11	0.57	0.467	0.08	139
42	GL 25A	2941	K1V+G	___	5.57	15.40	0.72	0.397	0.25	76
43	GL 334.2	44897	F9V	0.08	5.93	19.19	0.59	0.358	0.49	52
44	GL 356A	47080	G8IV-V	___	5.39	11.37	0.77	0.422	0.18	91
45	GL 376	49081	G1V	0.2	5.37	15.05	0.67	0.418	0.18	92
46	GL 484	62207	G0V	___	5.95	17.38	0.56	0.363	0.5	51
47	GL 564	72567	G2V	0.05	5.86	18.17	0.58	0.366	0.42	56
48	GL 596.1A	77052	G3V	0.08	5.86	14.66	0.68	0.370	0.42	56
49	GL 672	84862	G0V	-0.36	5.38	14.33	0.62	0.422	0.18	91
50	GL 759	95447	G8IVvar	0.4	5.16	15.18	0.76	0.425	0.12	114
51	GL 779	98819	G1V	0.05	5.79	17.77	0.60	0.371	0.37	61
52	GL 788	100017	G3V	-0.09	5.92	17.57	0.59	0.364	0.47	53
53	GL 67	7918	G2V	___	4.96	12.74	0.62	0.475	0.08	139

days, 16 stars have been selected to round out the first 22 months of the DRM (Table 3.3-3). These stars fill in the gaps when long slews are required between known Jupiter and exo-Earth candidate targets. The 16 stars have an average completeness of 40% for Jupiter-twins distributed as log of semi-major axis, in circular orbits, between 0.7 and 10 AU. For several of these targets, it will be possible to detect planets as small as Neptune, near the IWA.

The last column of Table 3.3-3 shows the spectral resolution that can be achieved at SNR $= 10$ with a 30-day integration assuming a Jupiter twin with 10^{-9} contrast relative to the host star. The minimum is $R > 50$. The detection times for three bands each with $R = 7$ and SNR $= 3$ are less than 1 day. Exo-S can quickly identify Jupiter twins and can spectroscopically characterize them to scientifically interesting resolution.

3.4 Observing Sequence

An observing sequence tool developed for the Exo-S DRM has been used to schedule observations of 53 targets in the first 22 months of the 3-year mission. All observations are made using either the 510–825 nm band, or for some of the Earth candidates, the 400–600 nm band with its 75 mas IWA. All 17

known Jupiter targets are characterized to $R > 50$ (Table 3.3-2). After the initial round of observations, follow-on observations would then be scheduled to perform background subtraction, confirm potential detections, and carry out characterization observations of newly discovered planets. Observations in the red band (600–1,000 nm) can also be scheduled for targets requiring characterization beyond 825 nm. Figure 3.4-1 shows the observing sequence in ecliptic coordinates. Targets are identified as Earth-like candidates (20 targets), known Jupiters (17), and Jupiter candidates (16).

The observations in Figure 3.4-1 are portioned as 451 days of slewing (average time 8.5 days/slew), 165 days integrating on target, and 35 days of overhead. For each target, overhead consists of 8 hours of setup and 8 hours of downlink. The observations require a total of 69 kg of propellant for a total ΔV of 2.2 km/s.

Solar-angle and Earth-angle pointing restrictions are included in the DRM. There are two solar restrictions: first, the Sun can be no closer than 30 degrees from the target. This is driven by diffraction of sunlight around the petal edges. A baffle on the front of the telescope is sized to allow pointing to this

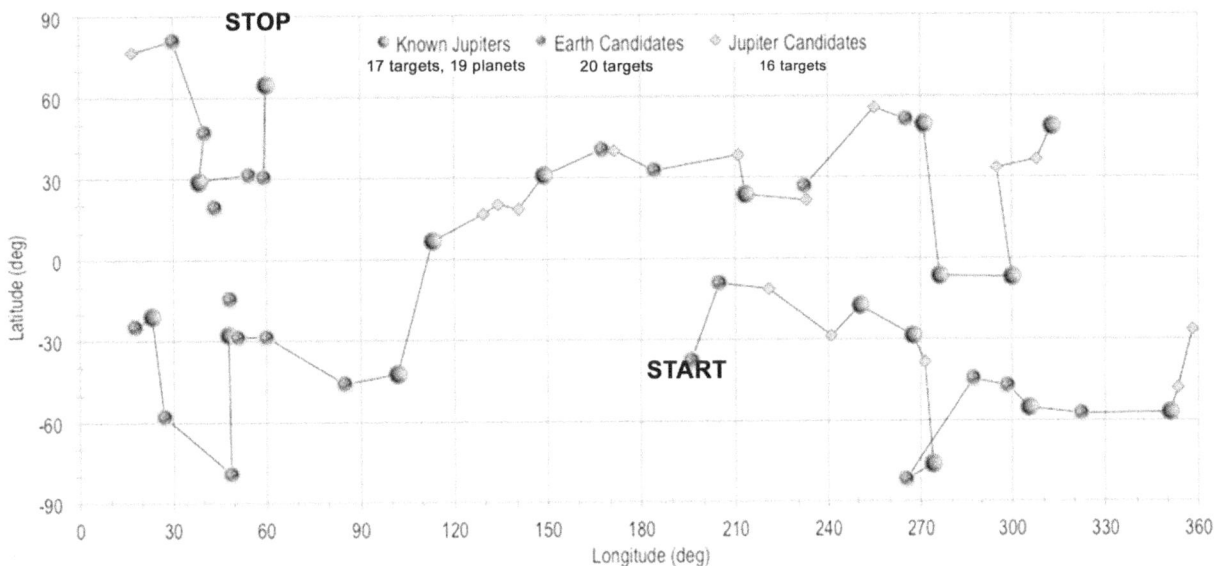

Figure 3.4-1. Observing sequence for the first 2 years of the mission. Arrows indicate progression of the observations in ecliptic longitude.

angle. Second, the Sun cannot illuminate the telescope-facing side of the occulter, and this is given a buffer of 7 degrees to allow for starshade misalignment and for structures extending in front of the plane of the starshade. This requirement restricts targets to ecliptic latitudes < 83 deg. For an Earth-leading orbit, the first year of the mission has a target-Earth angle constraint > 30 deg. The Sun and Earth constraints are centered on the ecliptic equator and limit observation duration and drive retargeting requirements for low-latitude stars. The DRM emphasizes high (>30 deg) latitude stars so that targeting and especially revisiting is not restricted by solar and Earth constraints.

3.5 Summary

In summary, the Exo-S probe's small IWA, high throughput, and efficient observing sequence enable a truly revolutionary scientific program that cannot be achieved from the ground. The Exo-S harvest includes (1) an Earth twin discovery program for the nearest and brightest Sun-like stars, (2) a 'guaranteed' science return through spectral characterization of known RV planets and mass determinations by resolving the *sin(i)* orbital inclination ambiguity, and (3) a Jupiter twin discovery program that takes

advantage of favorable targets along the starshade's path over the sky. For more than half of these Tier 3 targets, it will be possible to detect Neptune- and sub-Neptune–sized planets located within a few AU of the star. The mission's 95 mas IWA corresponds to an average separation of just 0.8 AU for the Earth twin sample, and to ~1.5 AU for the Jupiter-twin and known RV samples. The exposure times calculated for this DRM assume that each target has an exozodi surface brightness three times that of our own solar system.

For Earths in the habitable zone, the total observational completeness is 7.6 Hz distributed over 20 targets. Assuming η_{earth} = 0.22 the expected number of habitable Earth-like planets discovered is 2 in the first round of observation compared to a non-detection probability of 17%. For η_E = 0.3, the probability of detecting at least one Earth exceeds 90%. The Final Report will describe how the completeness and probability of detection increases with subsequent revisits in the final 14 months of the mission program. Additionally, the Final Report will also address the science potential in an extended mission, beyond the 36-month nominal lifetime of the Exo-S mission.

4 Baseline Design

> *A starshade enables the direct imaging of exoplanets with a probe-class mission using a small, conventional space telescope and a single launch vehicle.*

This section presents a strictly cost-driven design that is fully compliant with STDT guidelines. Except for the starshade, all flight elements are either flight proven, or space qualified and scheduled to fly by 2017. The observational performance detailed in Section 3 is capabilities driven and fully supported by this design.

Figure 4-1 shows the fully deployed spin-stabilized starshade spacecraft (left) and the 3-axis stabilized telescope spacecraft (right). Figure 4-2 shows the launch configuration with the telescope spacecraft stacked on top of the starshade spacecraft and fitting within a standard 5 m diameter launch fairing. The telescope is dedicated to exoplanet imaging and provides propulsion for retargeting maneuvers and formation control.

An alternate mission approach, to be detailed in the Final Report, is to operate the starshade as a secondary experiment with a multi-purpose telescope already planned for another purpose. The National Reconnaissance Office (NRO) telescope planned for WFIRST-AFTA (Wide-Field Infrared Survey Telescope–Astrophysics Focused Telescope Asset) is one example. In this case, the starshade spacecraft provides the propulsion for retargeting maneuvers and formation control. The starshade

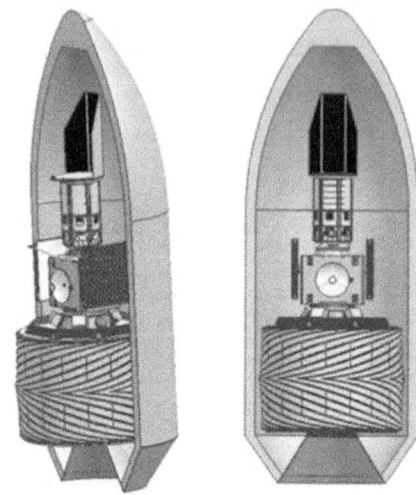

Figure 4-2. Starshade and telescope fit together in a standard 5-m-diameter launch fairing.

can launch separately and rendezvous with a telescope operating at Earth-Sun L2. The telescope spacecraft must launch with the equipment needed to operate with the starshade spacecraft. This consists of a fine guidance sensor (FGS) to sense the starshade position and a radio system for inter-spacecraft communications and range measurement.

4.1 Mission Design

Exo-S is a Class B mission with a 3-year prime mission duration. It launches from Cape Canaveral on an intermediate class expendable launch vehicle and operates in an Earth-leading orbit. Telecommunications and tracking are via the Deep Space Network (DSN), using 34 m-diameter radio antennas. Figure 4.1-1 shows the mission elements, interfaces, and nomenclature.

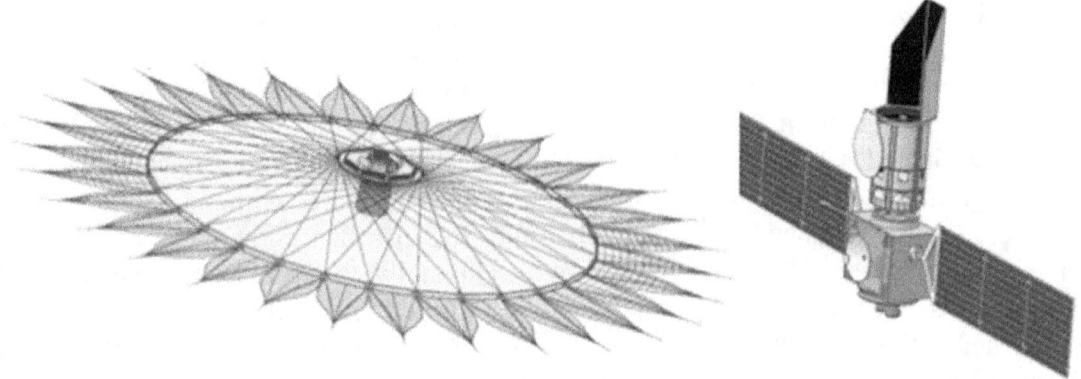

Figure 4-1. Fully deployed starshade spacecraft (left) and telescope spacecraft (right).

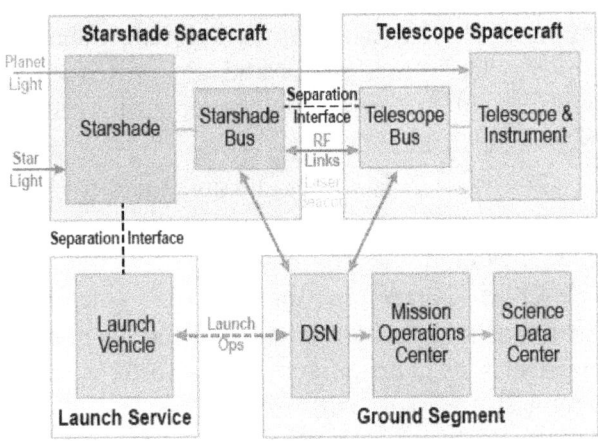

Figure 4.1-1. Exo-S mission interfaces with two spacecraft.

Table 4.1-1. System mass budget (kg) shows ample launch mass margin, far exceeding guidelines.

Element	Current Best Estimate	Contingency (%)	Max Expected
Telescope Spacecraft	**1,022**	**19**	**1,216**
Payload	**310**	**14**	**353**
Heritage Telescope	250	10	275
Sunshade	20	30	26
Instrument	40	30	52
Bus System	**560**	**21**	**678**
Heritage Bus	260	20	312
Solar Electric Propulsion	120	10	132
Solar Array	80	30	104
X-band, HGA, larger RWs	20	30	26
Additional Structure	80	30	104
Propellant	**152**	**22**	**185**
Xenon	100	—	120
Hydrazine	52	—	65
Starshade Spacecraft	**1,100**	**28**	**1,412**
Payload	**820**	**30**	**1,064**
Starshade	800	30	1,040
Launch Vehicle Interface	20	20	24
Bus System	**250**	**23**	**308**
Heritage Bus Avionics	110	20	132
Bus Structure	80	30	104
Solar Array	30	20	36
Propulsion	30	20	36
Propellant	**30**	**33**	**40**
Hydrazine	30	—	40
Total Launch Mass	**2,122**	**24**	**2,628**
Launch Capacity	3,550		3,550
Margin	1,428		922
Margin (%)	**67**		**35**
Margin Guidelines (%)	≥43		≥20

The launch vehicle deploys the two connected spacecraft on a direct trajectory to an Earth-leading orbit. The telescope spacecraft deploys its solar arrays and acquires a safe Sun-pointed state as the master spacecraft. Separation occurs after initial health checks and push-off springs provide a safe separation distance. The starshade spacecraft spins up, deploys the starshade, and acquires a safe, Sun-pointed state. The telescope spacecraft establishes formation at the specified separation distance and lines up on the first test target star and performance verification begins. Commissioning is complete within 90 days after launch and the prime mission begins.

Table 4.1-1 shows the system mass budget with a nominal launch mass estimate of 2,122 kg. The launch capacity is 3,550 kg and this gives a 67% launch margin, as compared to a minimum guideline of 43%. The maximum expected launch mass of 2,628 kg, also provides ample launch margin. Ample mass margins mitigate the cost risk of possibly needing the next larger and considerably more expensive launch vehicle.

The telescope spacecraft performs retargeting maneuvers using solar electric propulsion (SEP) and carries sufficient Xenon gas for a 5-year mission. Total SEP ΔV capacity is 5 km/s, including 2.2 km/s to perform the 22-month observing sequence detailed in Section 3. Subsequent observations focus on repeat visits, with longer observation times, and targets spaced farther apart, such that ΔV is accumulated at a much slower rate.

The telescope spacecraft also performs formation control using using chemical propulsion and carries sufficient hydrazine fuel for a 5-year mission. Total chemical ΔV capacity is 100 m/s.

The starshade spacecraft performs pointing and spin-rate control using chemical propulsion and carries sufficient hydrazine fuel for a 5-year mission. Nominally, no ΔV is required, but contingency propellant is carried for 30 m/s of ΔV capacity.

An S-band radio frequency (RF) link is maintained between the two spacecraft for both communications and the measurement of separation distance, via 2-way ranging. This is implemented with transponders and communication protocols developed by the Mars program. Both spacecraft have direct-to-Earth (DTE) links with 34-m DSN ground stations. Nominally, Earth communications are provided via the telescope, which relays commands and telemetry to the starshade. The telescope can store science data for up to 5 days and generally downlinks science data at the end of an observation via a high-gain antenna (HGA) at X-band.

4.2   Telescope Spacecraft

Figure 4.2-1 shows the fully deployed telescope spacecraft configuration. It consists of a telescope, instrument, and bus system.

4.2.1   Telescope

The baseline telescope is the commercially available NextView telescope with a 1.1-m aperture, designed for commercial Earth imaging. Four NextView telescopes are currently operational and a fifth is scheduled to launch in 2014. The heritage design is highly compatible with Exo-S requirements and minimal modification is needed.

A lightweight sunshade is added to allow pointing to within 28° of the Sun with no sunlight entering the telescope barrel. It

mounts to the existing hexagonally shaped cover door assembly and is sufficiently lightweight to avoid modification of existing telescope structures.

Figure 4.2-2 shows a ray trace of the telescope optics and the very small instrument optics behind the aft metering structure. The telescope barrel is the forward metering structure with spider mounts to the secondary mirror. A fast primary mirror affords a small secondary mirror with minimal blockage. The heritage secondary mirror assembly includes a five-degree of freedom (DOF) actuator. For Exo-S, this may be simplified to a single DOF actuator for focus adjustment only (piston).

4.2.2   Instrument

The Exo-S instrument supports all science observations detailed in Sections 2 and 3. It is simple, with all components flight-proven, and small, with overall dimensions of $30 \times 20 \times 10$ cm. It supports science observation in 3 overlapping wavelength bands (blue, green, and red) and 3 spectral resolution modes (full-band, 3-color, and high resolution). In full-band mode, it also supports 3 polarization states.

Table 4.2-1 details the 3 bands. Optical throughput varies between 40 and 60%, depending on the mode selected. The FGS is integrated as a separate channel, operating in parallel at wavelengths from 1.4–1.6 µm. Figure 4.2-3 shows the optical layout. Not

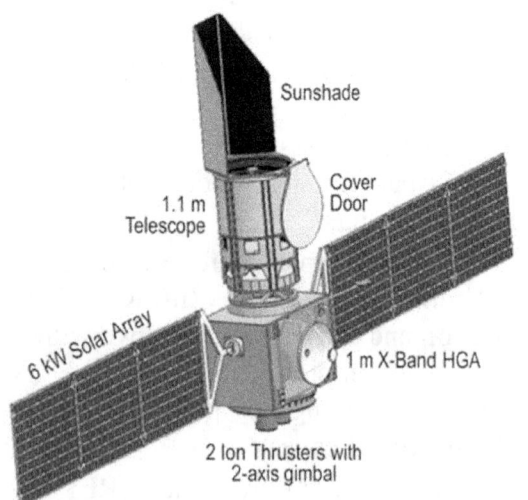

Figure 4.2-1. Deployed telescope spacecraft shows excellent match between heritage bus and telescope dimensions

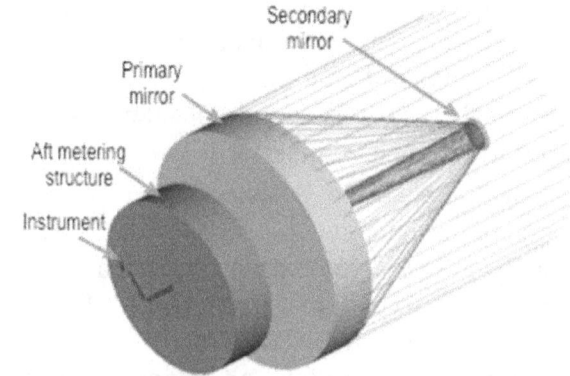

Figure 4.2-2. Optical ray trace shows compact size of the instrument behind the telescope aft metering structure.

Table 4.2-1. Science observing bands.

Parameters	Observing Bands		
	Blue	Green	Red
Wavelengths	400–630 nm	510–825 nm	600–1000 nm
Inner Working Angle	75 mas	95 mas	115 mas
Separation Distance	47 Mm	37 Mm	30 Mm

shown are local detector electronics and a separate instrument electronics unit.

A fast-steering mirror (FSM) picks off the beam from the telescope secondary mirror and performs multiple functions: static alignment, dynamic jitter control, target searches, and spectrometer slit alignment with candidate exoplanets. The FSM control loop is closed around the FGS and contained entirely within the instrument. Beam pointing is controlled relative to a designated point on the detector to within 75 mas (3-sigma) at frequencies below 1 Hz. In spectrometer mode, the designated detector position is matched to a candidate exoplanet position.

The heritage spacecraft bus independently controls bore-sight pointing within 9 arcsec (3-sigma) and jitter within 30 mas (3-sigma) at higher frequencies. These specifications are consistent with the heritage bus design.

A set of filter-wheel mounted bandpass filters selects one of three science bands. The planet camera receives reflected light and the spectrometer receives transmitted light.

Wavelengths longer than 1 μm are transmitted and delivered to the FGS. Each instrument band corresponds to a starshade bandpass that provides the requisite starlight suppression at a designated telescope-starshade separation distance (z), such that the product of λz is held constant. Inner working angle (IWA) varies as a function of separation distance.

On the reflected path, a slider mechanism carries a set of bandpass filters and polarizers to control planet camera sub-modes. A flat mirror is included for broadband observations within the selected band. Stacked dichroic filters split the selected band into 3 sub-bands for the purpose of measuring color to aid in disentangling candidate exoplanets from background noise and sibling planets. The dichroic stacks direct the 3 bands to different locations on the detector for simultaneous measurements. One of two Wollaston prism type polarizers can also be selected to aid in the identification of debris disks and separating their polarized signature from candidate exoplanets.

The planet camera has a 1 arcmin FOV and magnification is set for Nyquist sampling of the Airy disk at 500 nm. The detector is an e2V CCD-273, selected for its best available read noise performance of 3 e⁻ rms. It is in the process of being space qualified for the European Space Agency's (ESA) Euclid

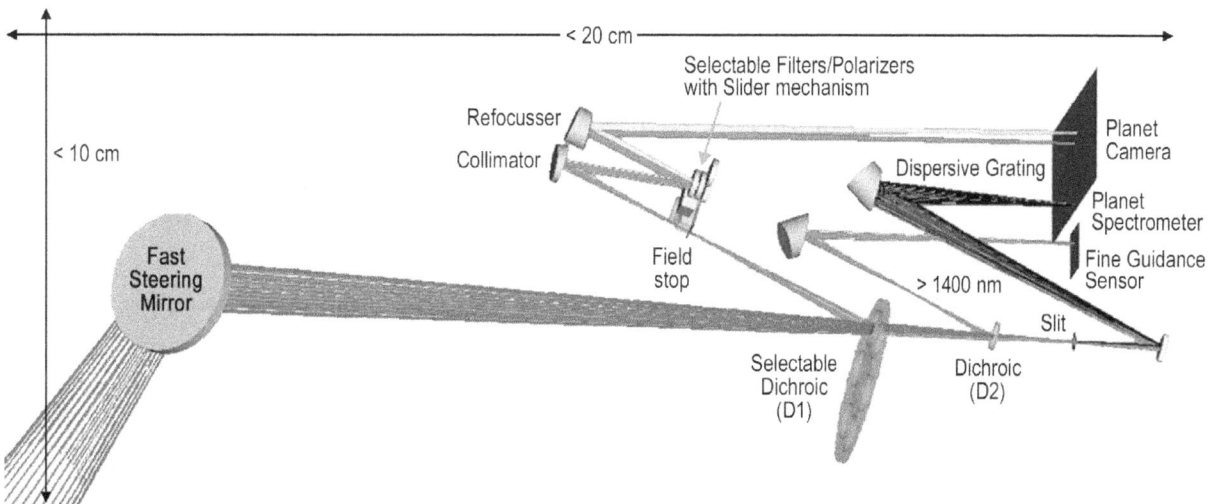

Figure 4.2-3. Science observing bands.

mission. Improved read noise will likely become available with other detectors already in development, but this is not assumed. The observation times allotted in the Section 3 DRM are based upon this existing detector. The standard format is 4K × 4K, but it is easily split into a 2K × 4K format, which still provides more than ample real estate. Part of the detector is unused and will be masked off. The detector is passively cooled to 153 K. Table 4.2-2 summarizes important detector parameters.

The planet spectrometer has a 0.2 × 2 arcsec FOV and magnification is set to match the spot size and pixel size at the longest wavelength. It shares the planet camera detector. The spectrometer is a simple Rowland design using a long slit and fixed diffraction grating to disperse planetary/debris disk spectra along a single row of detector pixels. Neighboring rows capture the background noise spectrum.

The grating is ruled to provide Nyquist sampling at the highest spectral resolution ($\lambda/\Delta\lambda$) of 200 and the shortest wavelength. On-chip binning provides lower spectral resolution options to limit integration times on a target-specific basis.

The FGS has a 2 arcmin FOV and magnification is set to spread the starshade point spread function (PSF) across several pixels to aid in centroiding. It operates at wavelengths between 1.4 and 1.6 µm. The detector is a Mercad (HgCdTe) type Hawaii H1RG detector, as used on the Hubble Space Telescope (HST) and detailed in Table 4.2-2. It installs on the same focal plane with the science camera detector and is passively cooled to 120 K. Operational use of the FGS is detailed in the following section.

4.2.3   Formation Sensing and Control

This section details the design for formation sensing, acquisition, and control. Axial and lateral components are sensed and controlled in distinctly different fashions and are discussed separately. The development of formation sensing and control algorithms is also discussed in Section 6 as a technology demonstration effort.

4.2.3.1   Lateral Control

During science observations, telescope lateral (cross-track) position is controlled to within ±1 m (3-sigma) to keep the telescope within the dark shadow created by the starshade. The FGS measures the relative position of the starshade and target star by simultaneously observing the starshade's point-source laser beacon and infrared starlight that diffracts into the shadow region. The error signal within the shadow is highly linear in power and enhanced spatially, as the long wavelength starlight appears as a PSF at the starshade tip.

Lateral position error is estimated to an accuracy of ±20 cm (3-sigma). Position is controlled with commercially available hydrazine thrusters capable of delivering small impulse bits. The environmental disturbance in Earth-leading orbit is very small and the control bandwidth is set by the thruster impulse bit and uncertainty in predicting lateral velocity. An estimated thruster firing period of >5 minutes improves sensor accuracy with long integration times.

4.2.3.2   Formation Acquisition

Figure 4.2-4 shows the steps by which lateral formation is acquired as part of each retargeting maneuver. The FGS is used in various modes throughout this process.

The transit phase of each retargeting maneuver consists of periods of acceleration, coasting, and deceleration. Propulsion is

Table 4.2-2. Detector parameters.

Parameters	Planet Camera & Spectrometer	Fine Guidance Sensor
Detector Type	E2V CCD-273	Hawaii H1RG
Format	2K × 4K	1K × 1K
Pixel Size	12 µm	15 µm
Field of View	1 arcmin	2 arcmin
In-band Quantum Eff.	≥70%	≥70%
Read Noise	≤3 e⁻ rms	≤30 e⁻ rms
Operating Temp	≤153K	≤120K
Dark Current	≤5.5x10⁻⁴ e⁻/pix/s	≤5x10⁻² e⁻/pix/s

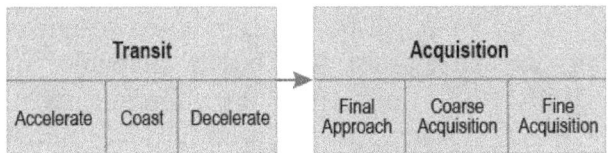

Figure 4.2-4. Retargeting and formation acquisition steps.

provided by ion engines axially aligned with the telescope, which precludes simultaneously pointing the telescope at the starshade to sense the laser beacon position. Thruster firings are controlled by accelerometers and the trajectory is periodically calibrated by stopping thruster firing, pointing the telescope at the starshade, and measuring the starshade laser beacon position relative to background stars. The transit phase delivers the telescope to within 10 km of final destination.

The acquisition phase begins with the FGS simultaneously capturing both the starshade beacon and target star, which improves the accuracy to which the trajectory can be projected. The approach period continues with ion propulsion and delivers the telescope to within 1 km of final destination with a small residual velocity. At this point, propulsion is switched to laterally firing hydrazine thrusters and the telescope continuously tracks the starshade beacon for further improvement in trajectory projection accuracy.

Coarse acquisition begins at about 30 m from final destination, as light from the target star is first perturbed by the starshade. The FGS senses the starshade beacon relative to the partially obscured target star. Non-linear diffraction effects limit sensor accuracy in this regime, but the trajectory is projected with sufficient accuracy to get within about 1.5 m of the final destination.

Fine acquisition begins with the telescope entering the dark shadow and the FGS performs as described at the beginning of this section.

4.2.3.3   Axial Control

The axial position, or separation distance, is loosely controlled to within ±250 km. Separation distance is sensed by the inter-spacecraft communications system using

2-way ranging. The Mars program has developed UHF frequency radios for orbiter/lander communications. They measure range to within about 10 m of accuracy. Exo-S range is greater than typical Mars applications and an upgrade to S-band is assumed, but this will be revisited in the Final Report.

The low disturbance environment in Earth-leading orbit allows the axial position to drift for weeks at a time without correction. Corrections are typically applied as part of retargeting maneuvers, using the efficient ion engines and firing at a small vector offset.

4.2.4   Telescope Bus

The telescope bus is an heritage design from an ideally analogous mission: ESA's PROBA-3 (PRoject for OnBoard Autonomy) mission, scheduled to launch in 2017. It operates in Earth orbit and will demonstrate the formation flying performance of a solar occulting spacecraft and a co-launched companion telescope spacecraft. The bus is procured directly from the vendor and not via an ESA partnership.

PROBA-3 is the fourth in a mission series that all use the same core bus architecture. Three are now operating in Earth orbit. PROBA-1 and -2 launched in 2001 and 2009, respectively, and have already far exceeded their design life. PROBA-V (V is for vegetation observations) launched in 2013 and is currently operational.

ESA classifies the PROBA missions as "small class" missions, which is most comparable to NASA Class C. However, the bus is fully redundant and ESA now has a single parts class for all of its missions. The redundancy and parts quality are fully compliant with a NASA Class B mission. Some parts, such as the main processor, are used on ESA "cornerstone" missions, which are comparable to NASA Class A.

The PROBA vendor has valuable experience with both formation flying and SEP. They supply both ion propulsion thrusters and integrated SEP subsystems. PROBA-2 used SEP with T5 ion thrusters developed by the same bus vendor. A larger T6 thruster is now space qualified, is a good match for Exo-S, and is adopted for the baseline design. It produces a maximum thrust of 154 mN with a specific impulse of 3,800 s and consumes 4.5 kW of power. All Exo-S SEP components (e.g., thrusters, tanks, power supplies, gimbals, etc.) are heritage from ESA's Bepi Columbo mission, which is scheduled to launch in 2016.

Figure 4.2-5 shows the bus layout and identifies the SEP components and interfaces to the telescope and starshade spacecraft. The heritage bus dimensions, construction, and general layout are preserved. The command and data handling (C&DH), power, and hydrazine propulsion subsystems are used as is. Key modifications are summarized as:

- Add a SEP system with gimbaled thrusters mounted below bus, fuel tank mounted inside bus, and power supplies mounted in compartment vacated by heritage instrument
- Add a 2-wing, 1-axis gimbaled solar array producing 6 kW at end of life (EOL)
- Add an X-band downlink with 1-m HGA and 6 W transmitter with PROBA-V heritage
- Beef up structure to carry additional mass
- Upgrade reaction wheels for greater momentum storage and torque

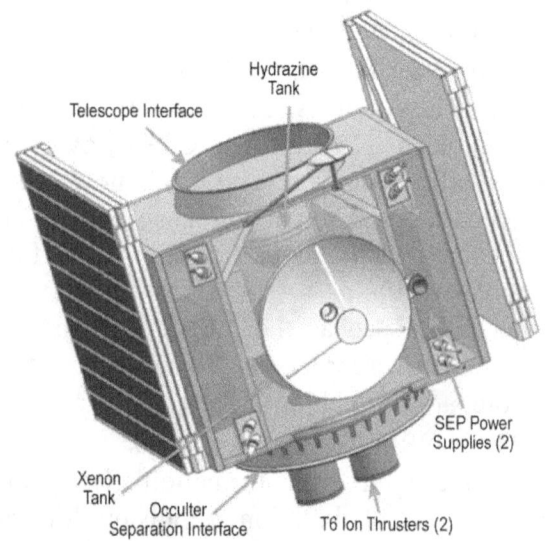

Figure 4.2-5. Telescope bus layout.

PROBA-3 is used in this baseline design as an existence-proof of busses capable of fulfilling this mission at an affordable cost. The actual spacecraft bus used for an eventual mission will be selected through a competitive process.

4.3 Starshade Spacecraft Design

This section details the starshade spacecraft optical, mechanical, and bus designs.

4.3.1 Starshade Optical Design

The starshade function is to block starlight and create a smooth apodization function to limit starlight diffracting into the shadow region. Ideally, the apodization function is a continuous gray-scale, but for the sake of a practical mechanical implementation, it is approximated as a binary function (all or none of the light passes at any point). This yields a central disk with flower-like petals extending radially from the disk perimeter. The apodization function is expressed as the open area as a function of radial distance from the center, or A(r). Figure 4.3-1 shows the baseline petal and starshade shapes.

A two-step optical design process is employed in iterative fashion to find an optimal solution. First, parametric studies are conducted based on a large number of approximate solutions and curve fitting to illustrate trends. Second, a select design is rigorously verified to provide the requisite starlight suppression at all points in the focal plane. Parameters are adjusted until the design is fully compliant.

Initial approximate solutions are generated using a linear program based optimization tool, following the approach first developed by Vanderbei et al. (2007). It finds the apodization function A(r) that minimizes the maximum bound on the electric field over the full shadow region and wavelength range, subject to the constraint that the electric field is everywhere less than the requisite suppression level. Computational efficiency is improved by constraining the real and imaginary parts of the electric field rather than intensity directly.

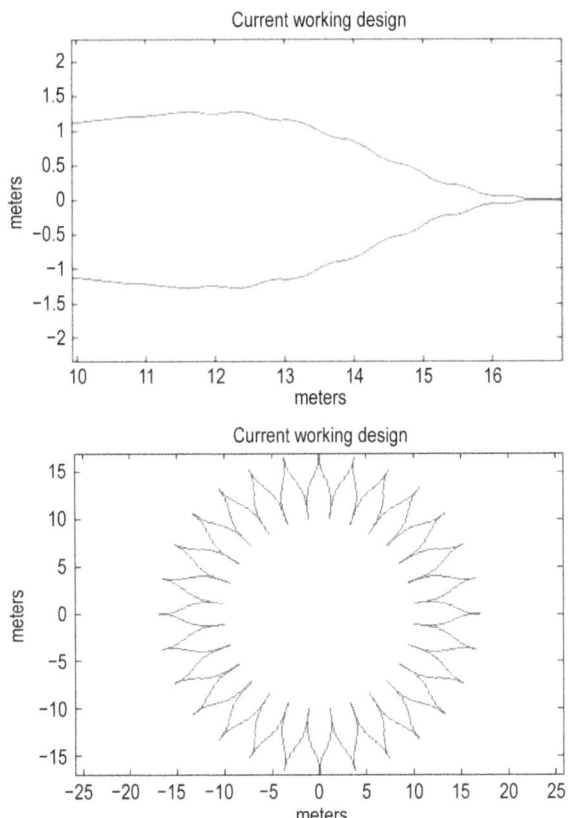

Figure 4.3-1. Exo-S petal and starshade shapes.

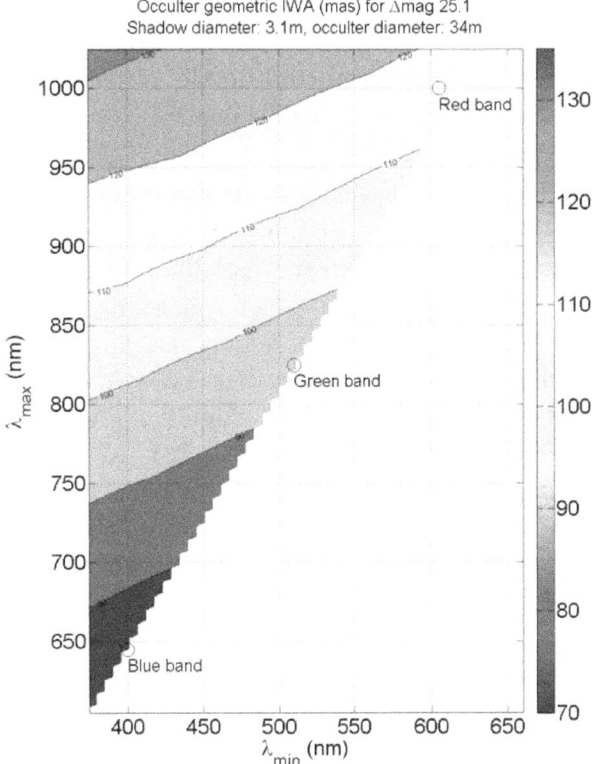

Figure 4.3-2. IWA capability varies as a function of upper and lower bandpass limits.

Practical implementation limits are set on starshade size (7-m petal length, 20-m inner disk diameter) and petal number (28). Additional constraints include a minimum petal tip width and inter-petal gap of 1 mm. The optimizer tool is used to define upper and lower wavelength limits as a function of IWA, as shown in Figure 4.3-2.

Specific point designs are further evaluated for science performance based on the combination of parameters. Planet sensitivity and IWA at a threshold of search completeness constrain a target list and observational performance is evaluated for this target list.

The baseline design is set for the observation of Earth twins at wavelengths between 510 and 825 nm (i.e., the green band) at 95 mas IWA and a separation distance of 37 Mm. Additional bands (blue: 400 to 630 nm and red: 600 to 1,000 nm) are available at inversely proportional separation distances and corresponding IWAs. Figure 4.3-2 illustrates

the IWA capability of the 3 selected bands. Each band provides identical optical performance at the designated separation distance and IWA.

4.3.2 Starshade Beacon

The starshade includes a laser beacon to support formation sensing as detailed above in Section 4.2.3.1. It is a conventional pumped-diode laser driving an erbium-doped fiber amplifier producing 2 W at 1,550 nm. It draws heritage from JPL's Optical Payload for Lasercomm Science (OPALS) mission, scheduled to launch in 2014. The beacon is mounted on the centerline of the starshade bottom deck, which faces the telescope.

4.3.3 Starshade Mechanical Design

The starshade is a large deployable structure that provides the requisite precision on-orbit shape. Figure 4.3-3 shows the fully deployed starshade configuration, consisting of an inner disk structure with 20-m diameter and 28 petals with 7-m length. The inner disk structure includes a fixed central hub structure and a deployable perimeter truss. All surfaces are covered with a loose-fitting opaque blanket.

The Exo-S starshade mechanical design represents a significant departure from earlier designs developed in the context of flagship missions. It stems from the 2010 Occulting Ozone Observatory (O_3) mission study (Savransky et al. 2010), which also targeted a total mission cost of less than $1B. A strictly cost-driven environment, combined with a rapid prototyping approach, led to quick definition of a simple and elegant design, which was prototyped with readily available and inexpensive materials. A full-scale petal prototype was deployment tested less than one year after formation of the O_3 study team. In most cases, the early prototype design remains the baseline flight design. Extensive heritage is leveraged from multiple flight-proven deployable antenna architectures.

The starshade mechanical architecture is largely driven by a few key guidelines:

1. The starshade is launched together with a telescope on a single low-cost launch vehicle with a standard 5-m launch fairing.
2. Petal construction is optimized for on-orbit thermal stability with near-zero CTE

(coefficient of thermal expansion) material controlling the critical width profile.
3. Petal construction is also optimized for stiffness, so as to not participate in low-order system modes and limit gravity sag during ground testing.

Guideline 1 led to a very compact stowed volume with perimeter truss and petals stowing concentrically around a central load-bearing cylinder (central hub), as shown in a cutaway view in Figure 4.3-4. Guideline 2 led to a very stable petal construction that is very stiff yet lightweight. Guideline 3 simplifies dynamic analyses and supports ground-based shape performance verification in a laboratory environment (1 g field).

Importantly, all starshade mechanical specifications are consistent with expected capabilities and the optical design is constrained by practical implementation limits. An inner disk diameter of 20 m matches the size of heritage antenna systems. A petal length of 7 m matches a coverage limit for common metrology systems. A petal quantity of 28 provides a favorable aspect ratio (length/width) for petal stiffness. Petal tip widths and inter-petal gaps are constrained to a minimum of 1 mm, to avoid manufacturing technology issues.

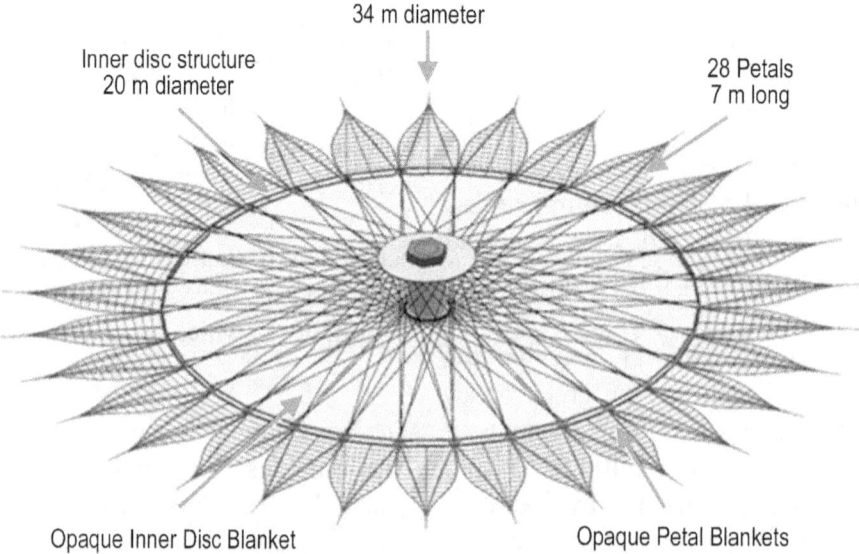

Figure 4.3-3. Fully deployed starshade configuration and major system elements.

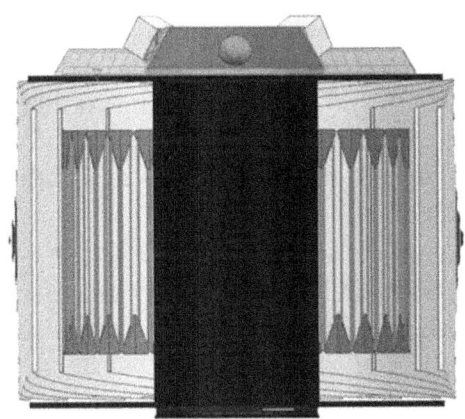

Figure 4.3-4. Cutaway view of stowed starshade.

Also important is an ample mass allocation, such that the design is not mass driven. The design is inherently lightweight, but this is driven primarily by stiffness requirements. Mass margin remains ample as the allocation reflects uncertainty in the blanket design, which represents a significant mass fraction. The current estimate of blanket mass is significantly less than the allocation.

The following sections detail the system configuration and each of the major elements: petals, inner disk, and blankets.

4.3.3.1 Starshade System Configuration

Figure 4.3-4 shows the stowed starshade configuration. It is short to fit within standard 5-m-diameter launch fairings with the telescope spacecraft stacked on top of it. It is axially symmetric to support spin-stabilization. It also provides the launch vehicle mechanical interface on the bottom surface.

The central hub structure includes a load carrying central cylinder with a 1.6-m-diameter to match standard launch adapters. The interior cylinder volume is left open, but the potential remains to install propellant tanks here. Top and bottom decks form a spool-like structure with the petals and perimeter truss stowed circumferentially around it. Separate space is allocated for stowing the folded up perimeter truss blanket. Petal blankets are fixed to petal structures and provide a degree of launch vibration damping.

Figure 4.3-5 illustrates the 2-stage starshade deployment process. The first deployment stage involves petal unfurling via the passive release of strain energy in a controlled fashion to avoid inter-petal contact. A precursor step is to release petal launch restraints. These are cords extending radially inward through interlocking features between overlapping petal center spines and are released by pyro-actuated mechanisms. Petal unfurling is then controlled by a motor-driven

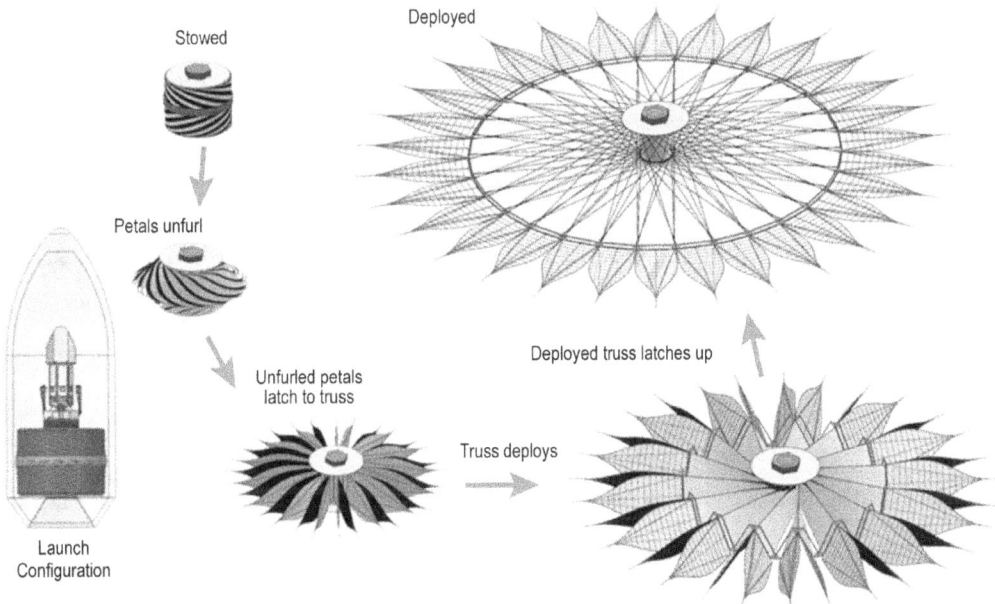

Figure 4.3-5. Starshade deployment sequence.

spool that slowly travels around the perimeter on a track as it unwraps a circumferential band holding the petals in place.

As each petal nears its fully unfurled state, passively actuated spring-loaded pop-up ribs rigidize the petal out of plane. Fully unfurled petals are vertically aligned per the inner disk longerons to which they are attached.

The second deployment stage is deployment of the perimeter truss. Two motor-driven spools draw in cables that snake through the perimeter truss tubes. This pulls the folded parallelogram-shaped bays into a horizontal position, which in turn pushes the perimeter bays and attached petals radially outward, while rotating the petals 90 degrees into the final starshade plane. High modulus flexible spokes precisely register the perimeter truss to the central hub and thus precisely position each petal.

The starshade surface presented to the telescope is less than 10% reflective to limit stray light from bright objects (e.g., Earth, Moon, Venus, Jupiter). Equipment extending out of the petal plane and toward the telescope is limited to preclude scattered sunlight, as illustrated in Figure 4.3-6, for the worst-case Sun position.

The extension of hardware out of the starshade plane is also restricted in the opposite direction (toward Sun and target star), but to a lesser extent. The objective is to limit solar shadows cast onto the petals, which contributes to non-uniform petal deformations.

The starshade bus system mounts directly to the central cylinder. One exception is the fixed solar array, which is mounted directly to the upper deck of the central hub structure. This gives the requisite cell area, which is not available on the bus structure.

Dynamic analyses via finite element structural models confirm the dynamics goals are achieved. The first elastic system mode is above 1 Hz and low-order system modes resemble typical Zernike patterns for a disk. The first petal mode is relatively high at around 5 Hz, such that the petal acts as a rigid body and does not participate in low order system modes. A spin-stabilized spacecraft further simplifies starshade dynamics. There are no reaction wheels, which eliminates a source of vibration. The telescope spacecraft performs formation control. Thruster firings on the starshade spacecraft are infrequent and never occur during science observations. The starshade spin-rate is specified to limit the wobble angle to a small fraction of the pointing error budget. On-orbit thermal stability performance is also excellent, as detailed in Section 6.

4.3.3.2 Petal Design

The petal, shown in Figure 4.3-7, is a lattice of graphite composite members called battens and longerons that intersect a longitudinal center spine. The lattice is highly mass efficient and yet very stiff in-plane. Flanking both sides of the battens along the longitude of the petal are

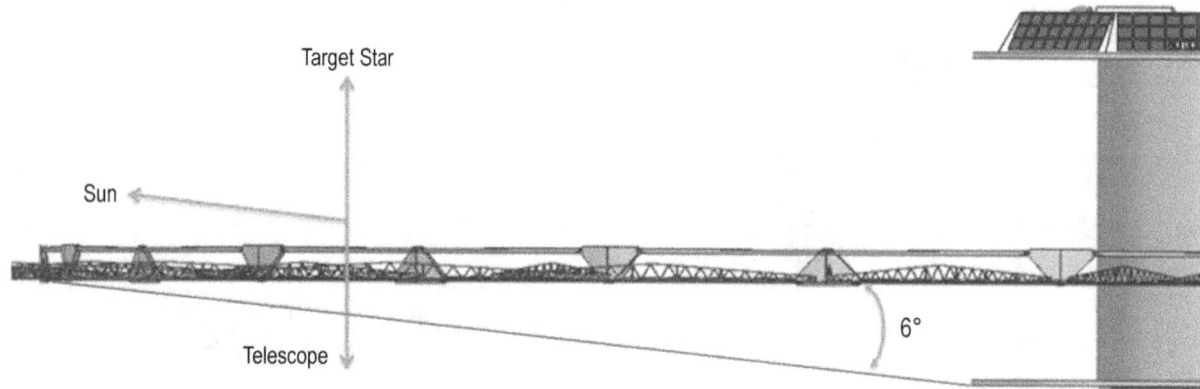

Figure 4.3-6. Geometry for avoiding scattered sunlight.

the structural edges, to which the separate optical edge segments attach.

The root of the petal is formed by the base spine, which is attached to both the center spine and structural edges, providing a rigid structure to which the petal can be attached to the truss. It includes two hinge points for the unfurling portion of deployment and two precise latches that position the deployed petal in-plane. Deployed out-of-plane petal stiffness with respect to the inner disk truss structure is achieved via the two base spine latches and root ends of the deployable ribs that latch to the inner disk truss to create a deep and consequently stiff beam connection of the petal to the truss.

The longerons provide in-plane shear stiffness. The longest longerons also serve as hinge pins for the passively deployed pop-up ribs that provide out of plane stiffness. The central spine is a foam core construction and stores most of the strain energy when petals are furled around the central hub.

The lateral batten members control and stabilize the critical petal width profile. They are constructed of carbon pultruded rods with a remarkably low CTE of -0.2 ppm/°C. The baseline material is a commercial product developed for sporting goods and is very low in cost. A further reduction in CTE to -0.1 ppm/°C is considered readily available, but not assumed here.

4.3.3.3 Inner Disk Design

The inner disk consists of a central hub and a perimeter truss with interconnecting bicycle-like spokes. This section further details the perimeter truss, as shown in Figure 4.3-8. Bays are constructed of longerons, diagonals, and battens. All members are made of a lightweight, near-zero CTE carbon fiber composite material for thermal stability. Key system trades are tied to the selection of bay member dimensions, as follows:

1. Longeron length, batten width and the number of bays/petals, sets the deployed diameter and petal width.
2. Longeron length and batten height sets the stowed starshade height and is constrained by the launch vehicle fairing.
3. Batten height also contributes to out-of-plane stiffness.

Batten members are dimensioned to provide an inner disk diameter of 20 m and 28 bays/petals.

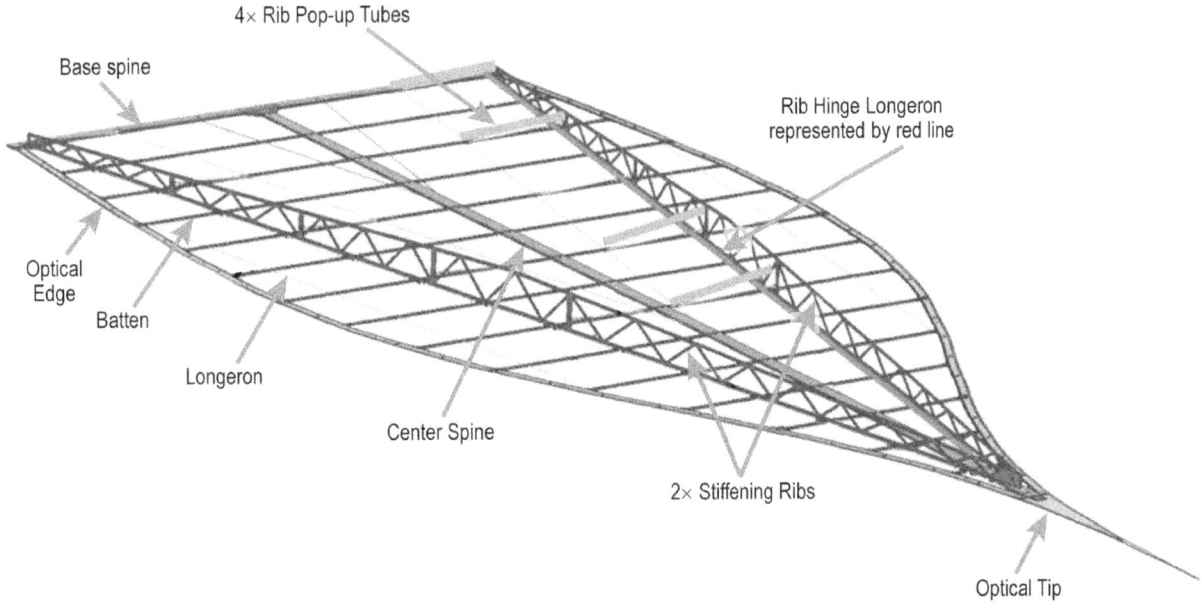

Figure 4.3-7. Petal construction.

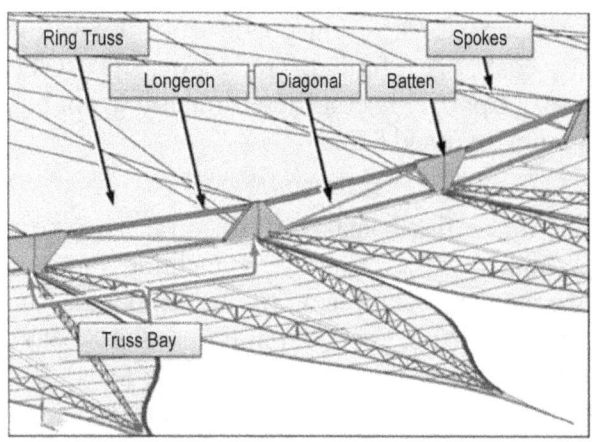

Figure 4.3-8. Perimeter truss of inner disk.

Petals are installed to perimeter truss longerons with two hinges. Shimming features are installed in both axial and tangential directions. Registration features are included on truss longerons and petal attachment points to enable precise petal positioning relative to truss node points (where forces are nulled), which are the most repeatable part of the truss, in terms of deployed position accuracy.

4.3.3.4 Blanket Design

Starshade blankets function to limit transmitted light, reflected light (telescope facing side), stray light from micrometeoroid holes, petal launch vibration, and thermal gradients. The baseline blanket construction, for both petals and inner disk, is two spaced layers of Black Kapton. Attachment and folding methods are different for petal blankets and inner disk blankets.

Black Kapton is highly opaque, to limit transmission, and non-reflective, to limit reflected stray light from bright bodies, such as Earth, Moon, Venus, and Jupiter. The spacing between layers limits the hole area that allows a light-path to the telescope. The two layers will stop some very small particles, but the integrated hole area of these particles is not large and the separation distance is not set for this purpose. The spacer material is baffled foam and also serves to limit sunlight transmitted through micrometeoroid holes after multiple reflections. The spacer is also springy

and provides launch vibration damping between overlapping petals.

The petal blanket is installed only on the side facing the telescope and contributes to petal temperature uniformity. Black Kapton closely matches the thermo-optical properties of graphite, with both having a solar absorptance to hemispherical emittance ratio close to one. This has the effect of heating the blanket-facing structure surfaces with close to the same heat flux as the Sun-facing side.

Petal blankets attach with pseudo-kinematic interfaces and are pre-crinkled to limit thermal shear loads. Inner disk blankets fold around the central hub in origami fashion and deploy along with the spokes.

4.3.4 Starshade Bus

The starshade bus is based on the same heritage as the telescope bus and will share the same core avionics. The most significant difference is the structure. It is reconfigured to a conical structure to provide an efficient load path from the telescope spacecraft to the starshade. The upper ring matches the heritage interface ring on the telescope bus and the lower ring matches the central cylinder of the starshade, which in turn matches the launch vehicle. Another change is to reconfigure the fixed solar array with sufficient area to accommodate the full range of sun angles. The solar array is now mounted to the starshade upper deck and is sized to provide 500 W of power at EOL.

Figure 4.3-9 shows the bus configuration after separation from the telescope spacecraft. Figure 4.3-10 shows the launch configuration with the telescope spacecraft still attached.

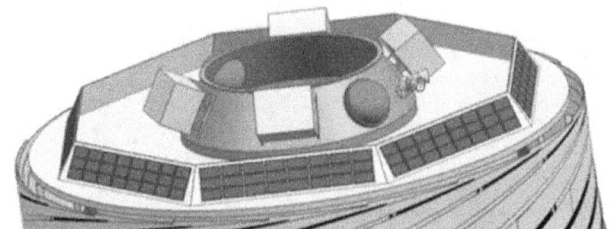

Figure 4.3-9. Starshade bus configuration, separated from telescope.

Starshade Bus

Figure 4.3-10. Starshade bus in launch configuration.

The starshade spacecraft is spin stabilized with significant simplification to attitude control (e.g., no reaction wheels). Pointing is loosely controlled to within 1° 3-sigma and a fraction of this allowable error is allocated to wobble, which results from imperfect mass properties. Spin-stabilization does require a different thruster layout with thruster clusters mounted on both top and bottom decks of the starshade so as to apply balanced torques with no net ΔV.

The spacecraft is repointed, or precessed, by firing thrusters to apply balanced torques in a direction 90° from the rotation. Repointing consumes propellant in proportion to the spin-rate. Selecting the spin-rate presents a trade between pointing propellant and the wobble angle, which is inversely proportional to the spin-rate squared. The baseline spin period is 3 minutes and this will be revisited in the Final Report.

The C&DH is a simplified version of the telescope bus C&DH with no science data storage and few operational modes.

The telecommunications is a simplified version with no X-band and no HGA.

5   Baseline Implementation

5.1   System Integration

The Exo-S integration and test (I&T) activities take advantage of clean payload interfaces and its dual spacecraft architecture to execute most tasks in parallel (see Figure 5.1-1). Experience on past missions has shown that this approach reduces schedule risk and allows a concentration of efforts to address unforeseen delays without affecting other parallel tasks. Parallel efforts begin with payload development, where the telescope, instrument, and starshade are built up separately, and continue into the two-spacecraft build-ups and payload integration.

Testing is part of a methodical system verification process. Requirements are identified and defined in Phases A and B, and validated through peer review and project-level review at the System Requirements Review (SRR) and Mission Definition Review (MDR). All requirements and interface definitions will be completed by the Preliminary Design Review (PDR). Each requirement is allocated to the appropriate element and designed to be verifiable, with a verification methodology assigned. Not all requirements will be verifiable by test (e.g., end-to-end system functional performance); some will be handled by analyses, modeling, or simulations.

The science instrument is integrated by an instrument partner and will include instrument-level ambient and cold functional tests including alignment, spectral response, distortion, and stray light rejection. The instrument will also be subjected to environmental testing including vibration, shock, and thermal cycling.

The commercial, off-the-shelf (COTS) NextView-like telescope will benefit from established I&T procedures and existing test facilities and support equipment. Assembly and test will be conducted at the vendor's facility. Testing will include environmental as well as performance tests.

Integration and testing of the instrument with the telescope (i.e., telescope payload integration) occurs at the telescope vendor's facility to take advantage of the test facilities and test equipment. Testing will include an end-to-end optical test with simulated target star and starshade laser beacon.

The vendor performing starshade payload I&T will have the requisite test facility for deployment testing of the starshade. Testing is

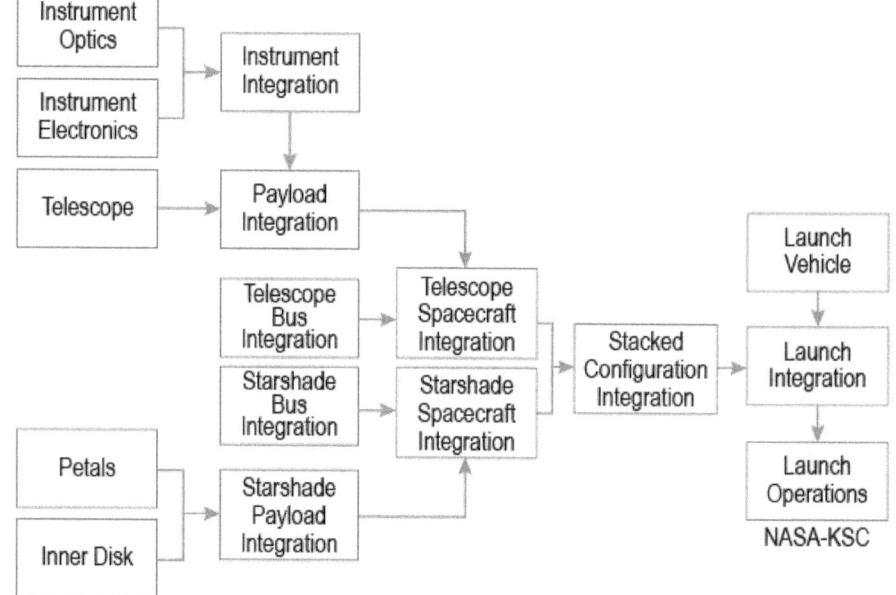

Figure 5.1-1. Exo-S integration flow chart.

analogous to large deployable antenna systems, requiring similar gravity compensation fixtures. The fully deployed starshade will require that the vendor has a test facility large enough to conduct deployment testing on the completed starshade. Most vendors involved in the design of large deployable antennas will have the required facilities, and stowed environmental testing and deployment testing will follow the methods developed for these antennas. Functional tests will include mechanical alignment of the deployed structure.

The formation flying (FF) hardware system on the telescope spacecraft consists of the fine guidance sensor (FGS), instrument electronics, hydrazine thrusters, and inter-spacecraft communications system including 2-way Doppler ranging. The FF hardware system on the starshade spacecraft consists of a laser beacon on the starshade payload and the equivalent Doppler-ranging radio system as on the telescope spacecraft. FF algorithms and software reside fully on the telescope spacecraft and are split between the instrument computer (lateral error prediction) and the spacecraft bus command and data handling (C&DH) computer (thruster control algorithms).

The FF actuators (hydrazine thrusters) are tested at the telescope bus level. The laser beacon is tested at the starshade payload level. The 2-way radio systems can be tested early as an integrated radio system and again at the integration and test of the two spacecraft busses.

The primary FF challenge for Exo-S lies with sensing lateral performance errors using the FGS. This involves sensing the starshade position as indicated by the laser beacon relative to: background stars (transit), unobstructed target stars (final approach), or obstructed target stars (coarse/fine acquisition and fine control). The FGS is an integral part of the science instrument and image processing is performed by the instrument computer.

FF sensing capabilities will be demonstrated in pre-Phase A. This includes optical performance verification of the starshade in the FGS band. A breadboard FGS and instrument computer will be tested in the subscale starshade test facility in development at Princeton University.

The FF algorithms and software, leveraging proto-flight software developed and ground-demonstrated for StarLight, Terrestrial Planet Finder Interferometer (TPF-I), and PROBA-3 (PRoject for OnBoard Autonomy-3) (Scharf et al. 2010), will be fully exercised in a Control Analysis Simulation Testbed (CAST) and a flight software (FSW) testbed. All functionality and performance requirements are first verified in CAST, then they are re-verified in the FSW testbed with the entire flight software suite (e.g., C&DH, telemetry, hardware managers, and fault protection) running on a flight processor in real-time. This software test and verification process—CAST then FSW testbed—has been followed successfully on missions such as the Mars Exploration Rover (MER) and Mars Science Laboratory (MSL).

Like most past NASA cost-capped competed missions, Exo-S plans to leverage existing commercial spacecraft designs to the greatest extent possible. Both spacecraft busses have identical core architectures, are produced by the same vendor and will be integrated and tested together prior to delivery.

A systems integrator will integrate each of the two busses with their respective payload and then integrate the two spacecraft systems into the stacked launch configuration.

The integrated telescope and starshade spacecraft are fully tested before shipment to the launch facility yielding high confidence for mission success. Testing includes: spacecraft performance testing, electromagnetic interference / electromagnetic compatibility (EMI/EMC), mass properties, vibration, acoustics, thermal vacuum and thermal balance, instrument functional testing, and Deep Space Network (DSN) compatibility testing.

Final operations at the launch site prepare the flight system for launch and mission operations. Preparations include flight system functional testing on arrival, Mission Operations Center (MOC) and Science Data Center (SDC) interface testing, integrating the two spacecraft on the launch vehicle, battery maintenance, fueling, final mass properties, and dynamic balance.

5.2 Mission Operations

Exo-S will observe a new target on average once every 12 days. Each target defines a repetitive operational cycle as illustrated in Figure 5.2-1. Science data is stored and downlinked at the end of each observation with the telescope spacecraft pointing its fixed antenna at the ground station. An operational sequence is then uplinked for the next target. Each sequence starts with a retargeting maneuver and formation acquisition.

Tracking and telecommunications are via the DSN using 34-m-diameter antennas. Engineering functions are via S-band and science data is collected via X-band. The two spacecraft have a separate S-band link for communications and 2-way ranging. Both spacecraft can communicate with the MOC directly via ground stations, but communications are primarily through the telescope spacecraft, which relays commands and telemetry to and from the starshade spacecraft. Navigational tracking requirements are typical of any single space telescope. Retargeting maneuvers and formation acquisition are performed in autonomous fashion without special DSN tracking or ground intervention. A link is maintained

during acquisition for monitoring purposes only.

The Exo-S ground system will follow the architecture developed for Kepler. The MOC will be at an organization familiar with operating Earth-orbiting and near-Earth missions. The spacecraft bus contractor will provide on-orbit subsystem technical support during the mission. DSN support is provided by JPL. Kepler ground system software will be modified for the mission-specific two spacecraft architecture of Exo-S. Science data will be archived at the mission's SDC and made available to the science community within a yet to be determined time of collection.

Science observations are precluded during specific seasonal meteor showers. The starshade will be oriented with its edge onto the meteor flux. The telescope spacecraft may still perform retargeting maneuvers during these showers. Science planning will account for these events to minimize any loss in observing time.

5.3 Mission Cost

The cost of the Exo-S concept is estimated at below $1B FY15 as required by the Probe Study Charter. This cost is achieved by leveraging existing industry capability wherever possible. The Interim Report baseline concept spacecraft busses are based on designs used by the European Space Agency (ESA) for their formation-flying solar occulting mission, PROBA-3.[1] The telescope is a high heritage COTS design like the 1.1-m NextView telescope.[2] The ground system and

---

[1] PROBA-3 is used in this baseline design as an existence-proof of busses capable of fulfilling this mission at an affordable cost. The actual spacecraft busses for an eventual mission would be decided through the standard procurement bid and proposal process.

[2] NextView is used in this baseline design as an existence-proof of a telescope capable of fulfilling this mission at an affordable cost. The actual telescope for an eventual mission would be decided through the standard procurement bid and proposal process.

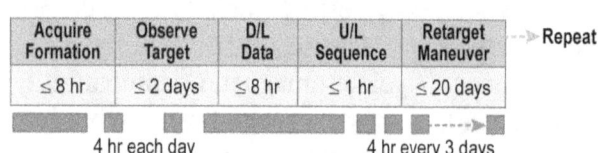

Acquire Formation	Observe Target	D/L Data	U/L Sequence	Retarget Maneuver	--▷ Repeat
≤ 8 hr	≤ 2 days	≤ 8 hr	≤ 1 hr	≤ 20 days	

4 hr each day   4 hr every 3 days

Figure 5.2-1. Operational cycle.

operations will follow the Kepler architecture. In parallel to the development of this report, Aerospace Corporation will produce an independent cost estimate through their Cost and Technical Evaluation (CATE) process. The CATE process endeavors to produce a low risk cost estimate and includes cost assessments for technical and programmatic risks; it was used as the cost evaluation method in the ASTRO 2010 Decadal Survey. To minimize differences between the Design Team estimate and the CATE estimate, two unofficial CATEs will be run before the final official CATE. This will afford the STDT the opportunity to adjust the baseline design should the CATE estimate come in at a higher number than the Design Team estimate.

All Design Team estimate costs are calculated in $FY15. The total mission cost includes all flight mission costs for Phases A–F, launch service costs, reserves, and all technology development costs following the start of Phase A. There is no contributed hardware or other support to the concept. The mission was assumed to be reliability Class B (per NPR 8705.4). The use of commercial busses was permitted for these studies.

The largest contributors to the total mission costs are the payload (starshade, telescope, and imager/spectrometer instrument), two spacecraft, launch services, and reserves. Launch services costs were specified by the charter guidelines. For the lowest cost intermediate class launch vehicle, the cost was set at $130M by study guidelines based on Team X launch vehicle cost data. Reserves were calculated at 30% of the total project cost excluding launch services costs with an additional 20% reserve for items requiring technology development (50% total). This exceeds the typical NASA proposal requirement of 25% cost reserves.

The starshade estimate is based on starshade technology expert opinion drawing from experience with large deployable antenna technology development efforts. A more detailed grassroots estimate and a quasi-grassroots cost model estimate are planned following this Interim Report.

The telescope and the imager/spectrometer costs were generated from two widely accepted statistical models, both utilizing only objective input parameters. The imager/spectrometer is modeled as an optical instrument using the NASA Instrument Cost Model (NICM), which is based on over 150 completed flight instruments. NICM is the primary NASA instrument cost estimation tool and has been in wide use for over 10 years. The telescope estimate is derived from aperture size and is calculated from statistical fits to historic actual costs given in "Update to Single-Variable Parametric Cost Models for Space Telescopes" (Stahl et al. 2013). The aperture-based model was chosen over the mass-driven model since, unlike mass, there is no uncertainty in the valuation of the telescope's aperture diameter. The Luedtke and Stahl telescope cost model predicts that a first unit telescope with a 1.1-m aperture would cost around $60M FY15. From the original 2012 paper (Luedtke and Stahl 2012), the authors establish a telescope cost for follow-on units at $C_n = C_1 * n^{-0.3}$, where C_1 is the first unit cost and C_n is the cost of the "nth" unit. Since Exo-S will be using the sixth build of the NextView telescope design, the model would estimate its cost at $35M FY15. Like this telescope model, NICM is also based on actual costs of flight hardware—both models represent the as-built costs with all reserves consumed. Since additional reserves are layered on top of these estimates in the total mission cost, they are both conservative in nature.

The telescope bus cost estimate is primarily based on the Aerospace Corporation's Small Satellite Cost Model 2010 (SSCM10). The Small Satellite Cost Model is an objective statistical model that estimates satellite bus costs at the subsystem level. The model was originally developed in 1995 to cost COTS busses but has been expanded to include many non-Earth NASA missions up to New

Frontiers–class missions. Today, SSCM10 can be used to estimate costs for busses with wet masses between 100 kg and 1,000 kg; both Exo-S busses fall within this range. The telescope bus includes a solar electric propulsion (SEP) subsystem that cannot be costed with the SSCM10. These costs were estimated by a JPL SEP expert with access to Team X hardware cost data, and were added to the SSCM10 estimate.

The starshade bus cost was also estimated using the SSCM10 cost model. Since it is largely the same design as the telescope bus (and both busses are expected to be built jointly by a single vendor as in the PROBA-3 and Deep Impact examples), only the recurring costs were applied.

SSCM10 also produces assembly, test, and launch operations (ATLO) costs as part of their model estimate. These costs were separated from both bus estimates and summed under Work Breakdown Structure (WBS) 10.0 as required in the NASA Standard WBS. Procurement overhead was applied to all bus and ATLO estimates as this work will be done by commercial vendors.

Exo-S draws heavily from Kepler for its ground system design. Exo-S has a similar heliocentric orbit like Kepler (though it is an Earth-leading orbit while Kepler is an Earth-trailing orbit). Like Kepler, the MOC is assumed to be run by a non-NASA contractor. The SDC will be the same as Kepler; only minor upgrades and setup costs are needed. The Design Team estimate for ground system development includes the Kepler actual costs as stated in the NASA CADRe (Cost Analysis Data Requirements) database and takes no credit for the reuse of their architecture. Post-launch operations cost will vary somewhat from Kepler since two spacecraft must be monitored. Kepler actual operations costs were used with their DSN costs replaced by a DSN cost estimate generated using the DSN Aperture Fee Tool developed by the Deep Space Mission System (DSMS) Plans and

Commitments Office. Other WBS costs (management, systems engineering, and mission assurance) are based on Team X models and are consistent with past JPL Discovery- and New Frontiers–class missions.

Technology development costs included in the Design Team estimate contain the costs needed to raise major concept elements requiring technology development from Technology Readiness Level (TRL) 5, at the start of FY17, to TRL 6. This work includes the development of a starshade system prototype and the development and test of algorithms for the spacecraft formation flying capability. The formation flying software development costs are derived from a similar JPL software development effort completed for PROBA-3 in 2010. The Exo-S Technology Development is discussed in detail in Section 6.

The Design Team is in the process of developing detailed estimates for elements of this concept requiring technology development since they contribute significantly to the total cost, and—as they are "developmental"—lack straightforward analogs from past flight missions. This work will follow this Interim Report.

In addition, and also following this report, Aerospace Corporation will deliver their first CATE for comparison to this estimate. After reconciliation to remove differences in assumptions between the CATE and the Design Team estimates, the costs will be compared WBS item by WBS item. If any significant differences are evident, the portion of the design driving the difference will be clearly identified in the comparison and the STDT and Design Team will evaluate the science impact of redesigning for lower cost. This process will be repeated again to identify any last disconnects between the CATE and the Design Team estimates. The design represented in the Final Report will include any adjustments needed to address these last issues and will be submitted to Aerospace

Corporation for the last and official CATE estimate.

5.4   Mission Schedule

The probe studies were directed to develop concept schedules based on a Phase A start at the beginning of FY17, project PDR in FY19, and a launch no later than December 31, 2024. Technologies requiring development must be at TRL 5 by the start of FY17 and TRL 6 at the start of FY19. Schedules are to include funded schedule reserves

The Exo-S schedule is shown in Figure 5.4-1. This schedule was developed from a combination of the Kepler schedule and the Deep Impact schedules including all technical delays and programmatically driven funding delays. Given that the basis for this schedule estimate is built from completed missions—with Kepler being the longest mission development of any Discovery or New Frontiers mission—and that margin is added on top of these completed actual durations, this schedule is a very conservative estimate. Kepler provides a good basis for the ground system and telescope, while Deep Impact captures the difficulties inherent in a two-spacecraft system development. The overall schedule (Phases A–F) is 128 months long and includes 92 months of spacecraft development (Phase A through launch and initial checkout), and 36 months of operations. Pre-Phase A technology development work on the starshade

and formation flying systems precedes the start of the project. The scope of this work is not included in the $1B cost cap and is discussed elsewhere in Section 6. In keeping with the study charter, Phase A begins at the start of FY17. Formulation (Phases A and B) runs for 35 months and includes requirements definition, system and subsystem design, and the start of procurements for long-lead items. It also encompasses the work needed to move the technology development of the starshade and formation flying systems from TRL 5 to TRL 6. The flight system implementation (Phases C and D) takes 57 months and includes the fabrication, integration, and test of the two flight systems. Implementation ends with the launch and initial on-orbit checkout in June 2024. The schedule shows an overall margin of 6 months, which is in keeping with JPL margin practices for a schedule of this duration.

The Phase A through Phase D duration for Exo-S is 92 months compared to 71 months for the dual-spacecraft Deep Impact mission, and 91 months for the Kepler exoplanet mission. The New Frontiers–class planetary missions are around the $1B cost cap placed on these studies and their Phases A through D schedules ran from 56 months (New Horizons) to 81 months (Juno), with an average of 71 months including the planned schedule for Osiris Rex (Origins Spectral Interpretation Resource Identification Security Regolith Explorer).

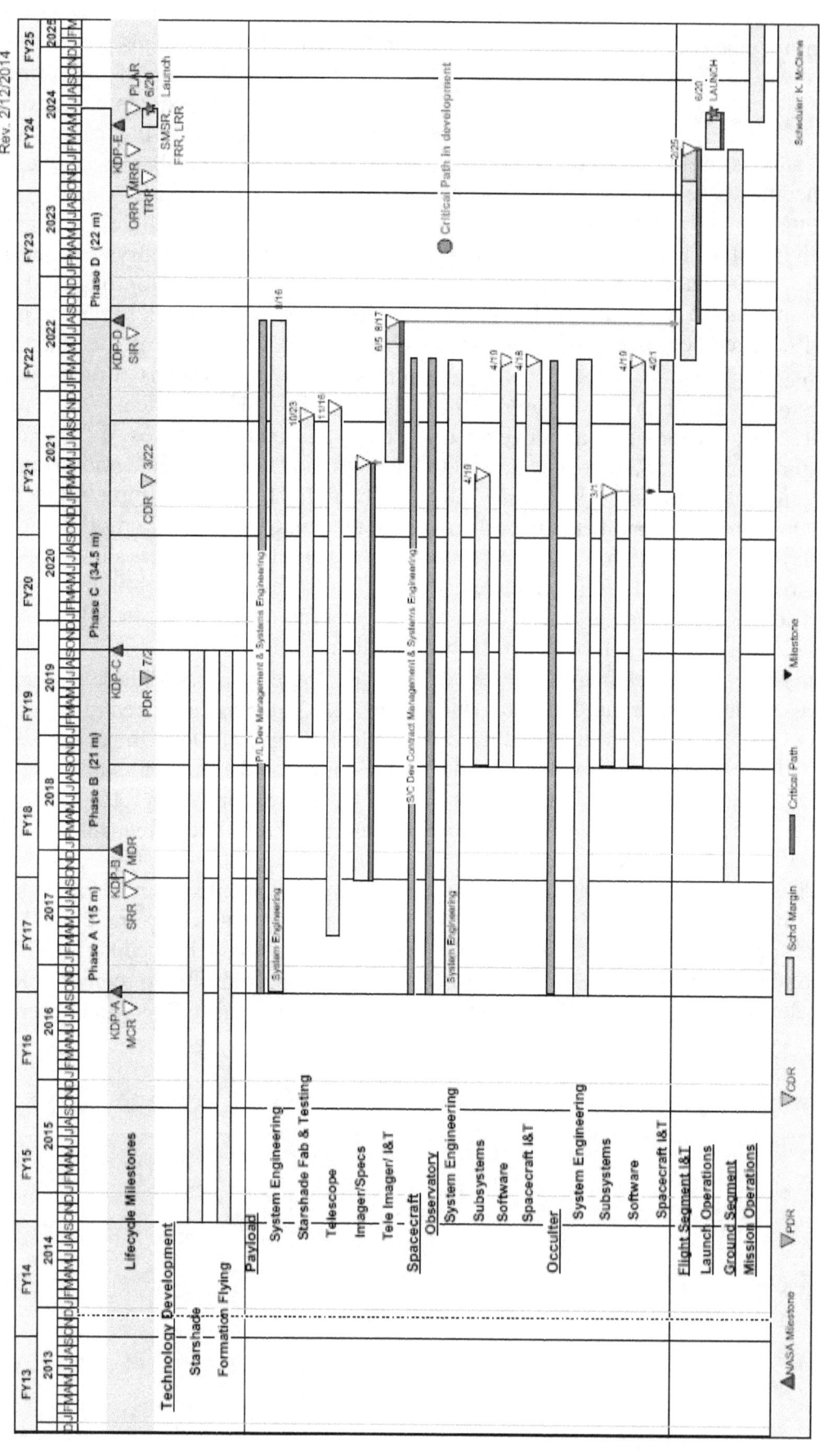

Figure 5.4-1. Exo-S top level schedule (preliminary).

6   Starshade Technology Readiness

Starshade requirements are well understood and considered imminently achievable. An ongoing NASA-funded technology program has already demonstrated key performance requirements, once considered tall tent poles, and efforts are underway to resolve any remaining tall-pole issues. A viable plan is presented for technology maturation consistent with STDT schedule guidelines.

The Exo-S mission uses a flight-proven telescope and instrument and soon to be flight-proven bus system; no technology development is required. All technology efforts are focused on the starshade and, to a lesser extent, formation flying (FF). The starshade draws heritage from flight-proven, large deployable antennas and, as such, its development risk is comparable, which is to say, manageable. FF control is straightforward, owing to a very benign disturbance environment and the challenge is focused on formation sensing from a long distance. This is manageable with use of the fine guidance sensor (FGS) integrated with the science instrument.

Figure 6-1 shows a flowchart and timeline of the starshade technology development plan. The timeline achieves Technology Readiness Level (TRL) 5 at the start of Phase A in 2017,

TRL 6 at the end of Phase B in 2020, and launch readiness by 2024. A number of critical technology issues are already resolved and funded efforts are underway to resolve most of the remaining critical issues. Additional efforts are proposed but not yet funded to resolve other issues (e.g., formation flying) and to develop the higher fidelity prototypes necessary to establish TRL 5 and TRL 6.

This section details the key requirements that drive technology issues, summarizes the completed efforts to resolve critical technology issues and, finally, presents plans to resolve the current open technology issues.

6.1   Requirements

Starshade optical performance is tied to four key requirements: deploy and maintain the correct shape; maintain sufficient opacity; fly in formation with the telescope; and mitigate solar glint.

6.1.1   Shape Requirements

The starshade shape and position requirements are established through two Technology Development for Exoplanet Missions (TDEM) studies and extensive error budget analyses. These requirements are consistent with

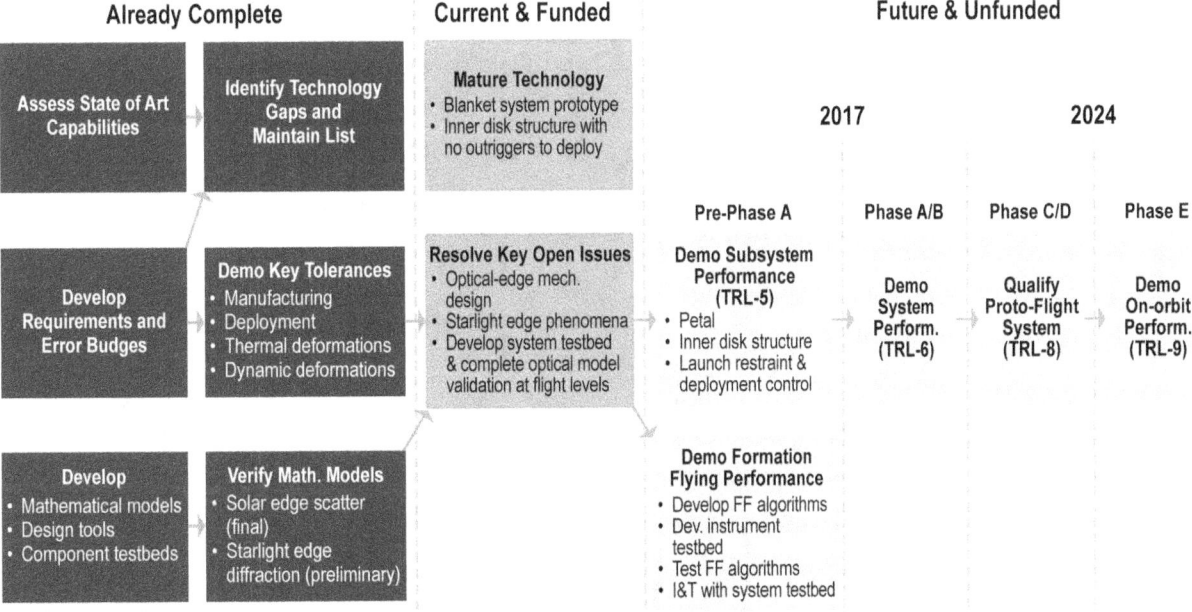

Figure 6-1. Technology development plan on-track to be ready for a new start in 2017.

achieving 1×10^{-10} planet contrast at the inner working angle (IWA). Completed TDEM studies include the development of a full-scale precision petal under TDEM-09 (Kasdin et al. 2012), and deployment of a 60% scale perimeter truss with 4 attached petals under TDEM-10 (Kasdin et al. 2013). Sections 6.2.1 and 6.2.2 describe the TDEM results, which, in both cases, were successfully consistent with starshades achieving better than 1e-10 contrast at the inner working angle.

The error budget analyses began with a joint effort between Northrop Grumman Aerospace Systems (NGAS), Ball Aerospace, and JPL to confirm consistency amongst several diffraction propagation codes (Shaklan et al. 2010; Glassman et al. 2010). The error budget was refined per knowledge gained from TDEM-09 (Shaklan et al. 2011).

Thermal requirements have been studied in detail with results reported in Shaklan et al. (2011). Note that uniform expansion of the entire starshade has no impact on performance. Uniform expansion is equivalent to moving the starshade closer or farther from the telescope and changes of even 1% have little impact. With a structural coefficient thermal expansion (CTE) of much less than 1 ppm/K, uniform expansion will be negligible even for extremes in solar illumination angle on the starshade.

Analysis of dynamics using finite element models of an earlier starshade design show that dynamics effects will be negligible; they are not addressed in this report but will be reanalyzed using a more detailed design in the Final Report.

6.1.2 Holes and Opacity

A cumulative pinhole area of $1\ cm^2$ is allocated for holes created by micrometeoroids and the associated contrast allocation is 1×10^{-12}. By comparison, a single 1-cm^2 hole leads to 3×10^{-12} contrast (Shaklan et al. 2010). This analysis assumed each pinhole is like an ideal aperture in a single-layer thin screen. However, the starshade will use two layers with cm-scale spacing between layers. If a micrometeoroid were to puncture all layers, the result would be a series of pinholes illuminated by other pinholes. Even if all the holes were aligned toward the telescope, the multiple scatter reduces the transmitted field strength at each layer, and also scrambles the phase of the final transmitted field. This will cancel the leakage fields at the telescope more effectively than direct transmission distributed across the shade. Thus, the tolerances outlined above are conservative, and can probably be relaxed after further analysis. Modeling of the integrated micrometeoroid flux shows that even for a single layer blanket, the $\leq 1\ cm^2$ hole area allocation is satisfied. (Arenberg et al. 2007). However, this does not account for seasonal micrometeoroid showers when the flux increases by a couple of orders of magnitude. A couple of times a year it is necessary to turn the starshade edge onto the shower for a period of 1 or 2 weeks.

The pinhole model is useful for placing a requirement on the average starshade transmission. The starshade blanketing material is highly opaque and not optically flat, such that residual light transmitted through the blankets in one part of the starshade will not interfere coherently with light passing through any other region. It then acts like a broad-angle scatterer, no different from distributed pinholes. The transmission requirement is then 1×10^{-7} and equivalent to $1\ cm^2$ over the 34-m-diameter starshade.

6.1.3 Formation Flying

The starshade is designed to produce a dark shadow that extends radially 1 m beyond the telescope aperture. Contrast degrades rapidly beyond the 1-m specification, as shown in Figure 6.1-1. Formation control is required to keep the starshade center positioned laterally within ±1 m of the telescope boresight. This requires sensing the lateral position error to within about 20 cm. The technology plans for demonstrating this capability are detailed in Section 6.3.3. The axial separation distance

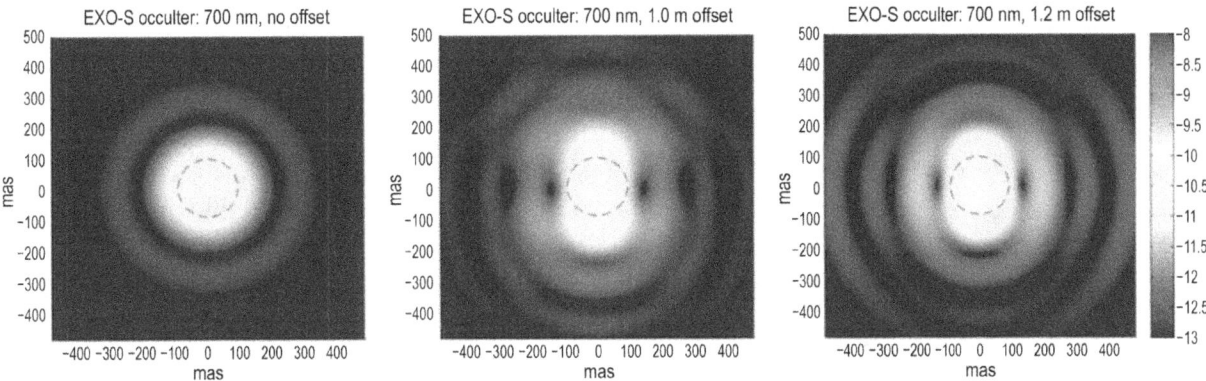

Figure 6.1-1. Image plane contrast at 700 nm with no lateral error (left), 1 m error (center), and 1.2 m error (right).

between starshade and telescope is loosely controlled to within ±250 km.

6.1.4 Optical Edge Scatter

Starshade optical edges will scatter and diffract a small fraction of sunlight and a small fraction of the target star light will enter the telescope. Section 6.3.1 details the modeling of this "edge-scattered sunlight" and the validation of that model. The allocated post-calibration contrast is 1×10^{-11} and this translates to a requirement that the scattered and diffracted sunlight should be equivalent to that scattered from a common razor blade. For safety reasons, and to reduce diffraction, the edge will be considerably more obtuse than a razor blade. If the edge has lower reflectivity than a razor blade, it can afford a compensating relaxation of its radius of curvature (RoC).

In addition, the optical edge must accommodate bending strain, associated with petal stowing, and thermal strain, associated with any mismatch in material CTE relative to the petal structure.

None of these requirements are individually difficult to achieve. In combination, however, they present a moderate material design challenge. For example, the TDEM-09 petal included graphite optical edges (same material as substrate structure) that satisfy all requirements, except for RoC. Subsequent RoC testing of many different types of graphite revealed that graphite was not likely to be a viable material.

The optical edge mechanical design is included on the list of current tall-pole issues and addressed further in Section 6.3.1.4.

6.2 Resolved Technology Issues

6.2.1 Manufactured Shape

The petal width profile must be manufactured to within a tolerance of ±100 μm. Compliance is demonstrated by test through a TDEM activity (TDEM-09) led by Professor N. Jeremy Kasdin of Princeton University.

Figure 6.2-1 shows the 6-m-long TDEM-09 petal prototype of graphite construction (1-m tip section not shown). By comparison, the baseline petal is 7-m long. Optical edge segments of matching graphite construction are precisely positioned and bonded in place to define the petal width profile.

The petal structure was assembled in a multi-step process. It was populated with metrology targets and precisely measured using a *large off-site* coordinate measuring machine (CMM) with ±5 μm accuracy over the full petal length. This knowledge was used to precisely position optical edge segments relative to local metrology targets on the structure, using a *small on-site* CMM with ±10 μm accuracy over a few centimeters. After bonding all 10 optical edge segments in place, the petal was measured a final time with the large CMM.

Figure 6.2-2 shows resultant edge position errors relative to a best-fit nominal shape. The edge profile is within tolerance over 99% of

Figure 6.2-1. TDEM-09 petal prototype used to demonstrate manufacturing tolerance on petal width profile. Micrometer stages for positioning edge segments shown at bottom right.

edge length. Optical performance was simulated, in terms of image plane contrast, by randomizing these single petal results over a full complement of petals. Figure 6.2-3 shows the results of this simulation expressed as a contrast probability distribution with a peak at the allocation of 2×10^{-11}.

TDEM-09 results fully demonstrate the achievability of the allocated manufacturing tolerances on petal width profile. The flight

build will benefit from investment in an *in-situ metrology tool*. This tool will be mated to the assembly table (i.e., optical bench) and used for petal assembly, edge installation, and final shape measurement without moving the petal. Improvement is also available in the optical edge machining accuracy, relative to the conventional CNC router used for this TDEM.

One simplification for this TDEM is the use of square-cut optical edge segments. The flight unit requires a sharp bevel cut edge to limit scattered sunlight. This may change the type of metrology sensor head, but does not invalidate the results.

6.2.2 Deployment Errors

Each petal attaches to the inner disk at two hinge points and the deployed position of these hinge points must be precisely controlled. The diameter of the best-fit circle through all petal hinge points represents the achieved inner disk diameter and the allocated tolerance (i.e., mean radial bias error) is ±0.25 mm. The random tolerance is ±0.5 mm in each of radial and tangential directions. There is by definition no tangential mean position bias. Compliance is demonstrated by test as a TDEM activity (TDEM-10) led by Professor N. Jeremy Kasdin of Princeton University.

3-σ error bounds for petal edge deviations (± 100 μm)

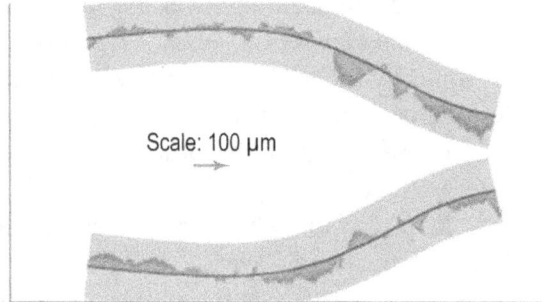

Scale: 100 μm

Figure 6.2-2. Measured petal shape error (green arrows) vs. 100 μm tolerance for 1×10^{-10} imaging (gray band) shows full compliance with the allocated tolerance.

PDFs for contrast with representative errors on each petal

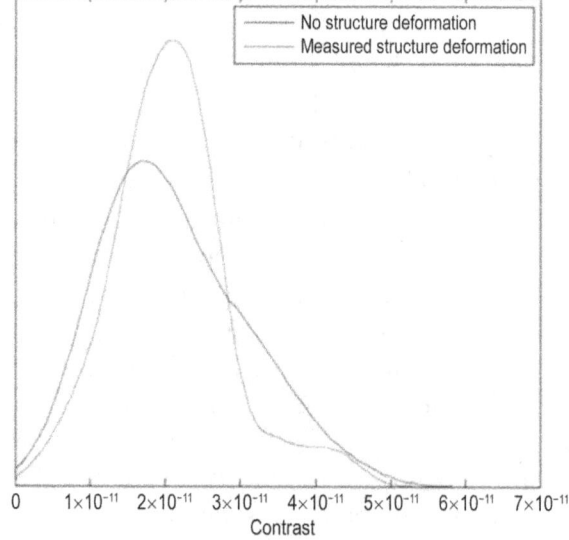

Figure 6.2-3. Contrast power density distribution per Monte-Carlo simulation of randomized errors on all petals.

Figure 6.2-4 shows the subscale partial system prototype developed for TDEM-10. It consists of: 1) 3-m diameter central hub of aluminum construction; an existing Astromesh antenna prototype with 12-m diameter; and 3) four petals 4.25-m long of mixed aluminum and composite construction. The existing Astromesh antenna was modified to add petal attachment fixtures and replace antenna webbing with spokes. The petals were sized to match the existing inner disk structure. Integration and test was performed at the NGAS Astromesh production facility in Goleta, California. Existing gravity compensation fixtures were used for the inner disk (see fan-shaped rails in Figure 6.2-3). Additional rails were added for the petals. Numerous optical targets were distributed around the prototype, but the final measurements were largely based upon targets closest to the hinge points. After each deployment, target positions were measured with both photogrammetry and a laser tracker.

After 10 initial deployment/metrology cycles, mechanical shims were installed to reduce the mean radial bias error. Additional shim cycles were not considered due to constraints on facility access. An additional 15-deployment/metrology cycles formed the final data set. Figure 6.2-5 shows the results, expressed as 90% confidence intervals with 3-sigma uncertainty. Small number statistics yield intervals that were effectively at the 4-sigma level. All intervals are contained within the allocated tolerance.

Radial errors show a residual bias that could be reduced with additional shimming. Tangential errors are minimal for the two inner petals and larger, but still within the tolerance limit, for the two outer petals. This behavior is an expected manifestation of the preexisting hardware. The petals need to be registered to truss node points at the junction between bays and where all forces are nominally nulled. These are points on the truss with precise deployment repeatability.

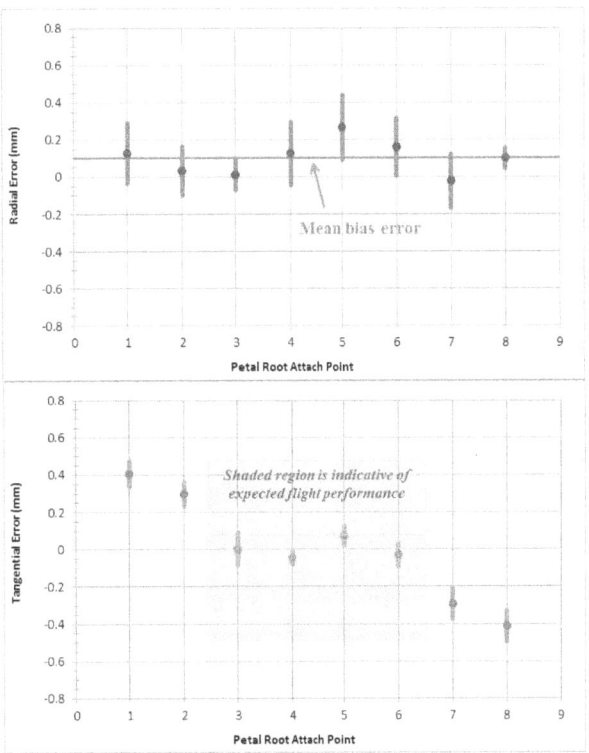

Figure 6.2-4. Deployed position tolerance demonstration. Petal root positions are measured after each of 20 deployments.

Figure 6.2-5. Measured deployment errors (3-sigma with 90% confidence) are all within tolerance allocations.

The existing Astromesh antenna provides no registration features to precisely locate the nodal positions. A retrofitted registration feature was possible for the primary nodes, but not the alternating slave nodes. A registration tool was installed to the primary node between petals 2 and 3 (*attach points 4 and 5*). A precision tool was used to locate attach points 3 and 6. Attach points on petals 1 and 2 (*attach points 1, 2, 7, and 8*) were positioned with further extrapolation and the errors started compounding. Future custom designs will include the necessary registration features at every node.

Future custom designs will also provide additional torsional stiffness, sufficient to avoid the need for complex outrigger struts to register the petal in plane. The existing Astromesh antenna requires adding separately deployed outrigger struts to register petals within the starshade plane. An acceptable outrigger deployment solution has not emerged and the gravity compensation fixture provided this function for the TDEM-2 activity. A significant deformation of truss shape occurred as a result. A new inner disk design is under development that provides the requisite torsional stiffness with a small strut at the petal root attaching to truss.

6.3   Unresolved Technology Issues

This section details three unresolved technology issues: solar glint, starlight diffraction, and formation flying. Solar glint requires demonstrating performance of an optical edge design. Starlight diffraction requires developing a subscale optical testbed to closely match flight-like diffraction behavior and validate the diffraction models. Formation flying requires demonstrating formation sensing performance and control algorithms in the subscale optical testbed.

This section does not address the additional technology efforts to mature technology with higher fidelity prototypes at increasing levels of integration. Significant engineering challenges are associated with these efforts, but they are arguably not considered unresolved technology issues.

6.3.1   Solar Glint

Exo-S observes target stars when they are positioned between 30° and 83° from the Sun. Equivalently, sunlight is incident on the starshade at 30° to 83° from surface normal and always on the anti-telescope side. A small but significant fraction of incident sunlight reflects and diffracts from the starshade optical edge into the telescope to appear as solar glint and contributes to instrument background noise. Specular reflection and diffraction of concern is limited to portions of the edge that are oriented normal to the Sun–starshade–telescope plane, as shown in Figure 6.3-1. Diffuse reflections of concern originate from a large fraction of the starshade edge, but are spread out over a full hemisphere.

This section details the modeling of solar glint, the flight design approach, and the optical edge mechanical design status, and technology development plan. A TDEM-12 activity to make headway in this plan is in progress and led by Suzi Casement of NGAS.

6.3.1.1   Modeling and Predicts

A starshade system model was developed to predict solar glint fluxes as a function of solar

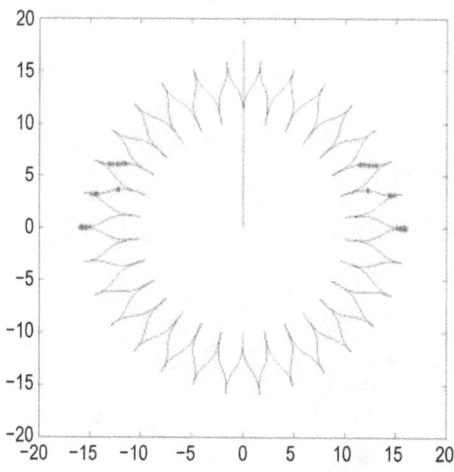

Figure 6.3-1. Lit-up edge regions. Red symbols indicate where specular reflected and diffracted sunlight originates. Sun is 10° into paper at top of figure (80° solar incidence). Units are meters.

incidence angle. The model was validated by testing a variety of representative edges in a scatterometer testbed, developed for this purpose (Martin et al. 2013).

Figure 6.3-2 compares model predicts to measurements of a commercial stainless steel razor blade. The model is in excellent agreement with measurements over solar incidence angles between about 50° and 80°. The testbed will be upgraded to improve sensitivity over the full range of Sun angles (30° to 83°). Figure 6.3-2 also shows that diffracted sunlight is the dominant term and the sum of all reflected sunlight is at least 1 visual magnitude dimmer than diffracted light.

The tested razor blade was representative and not intended as a flight solution. However, it accurately represents solar diffraction, which is independent of edge RoC and reflectivity (R). The reflected flux for other edge designs can be scaled in proportion to the product of RoC and R (i.e., edge surface area). The tested razor blade had a 0.2-μm RoC and was highly specular with 60% reflectivity. Any similarly specular edge with a RoC × R product of 12 will reflect the same solar flux into the telescope.

Figure 6.3-3 shows how solar glint influences image plane contrast at 80° solar incidence. The lobes correlate to edge orientations where diffraction and specular

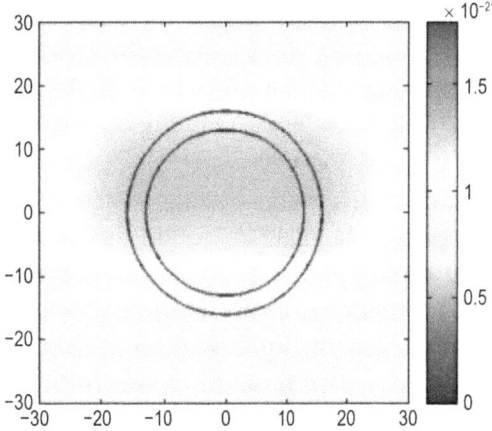

Figure 6.3-3. Solar glint contribution to instrument contrast (pre-calibration). Lobes correspond to lit-up edge regions.

reflection occur (see Figure 6.3-1). The lobe brightness is equivalent to a point source at 28 visual magnitudes and increases to 27 visual magnitudes at 40° solar incidence.

6.3.1.2 Flight Design Approach

The flight design approach is twofold. First, reflected solar glint is limited to 1 visual magnitude dimmer than diffracted solar glint. As in the test case, this corresponds to a RoC × R product ≤ 12. The current goal is a RoC ≤ 1 μm and reflectivity ≤ 10%.

Second, solar glint is calibrated to 1% of predicted flux, such that the systematic noise floor is limited to 1/5th as bright as the faintest exo-Earth in the Design Reference Mission (DRM). Solar glint is highly stable and can be calibrated as a function of solar incidence during long retargeting coast periods when the starshade and telescope can point at each other. A target star is not necessary for these calibrations.

6.3.1.3 Optical Edge Development

The mechanical design of the optical edge is challenged to provide the requisite RoC and reflectivity, while also accommodating the bending strain associated with petal stowing and any thermal strain associated with any CTE mismatch with the petal structure to which it is bonded. Graphite was the initial material of choice, as it matches the petal structure CTE

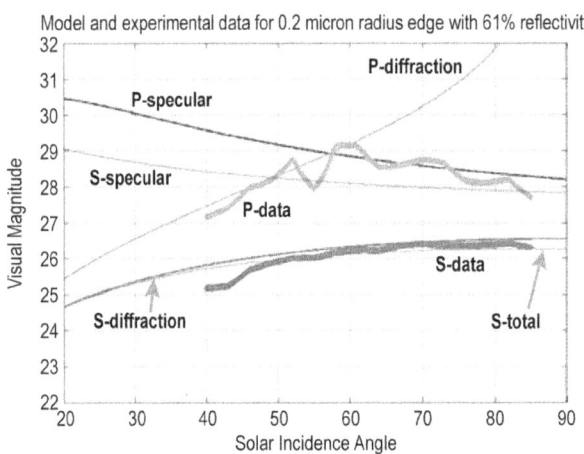

Figure 6.3-2. Model predicts compared to stainless steel razor blade measurements (not baseline design).

and has high strain capacity. Experimentation with a variety of graphite types and machining techniques showed that graphite does not satisfy the RoC requirement. A material that provides a sharper edge is required.

Multiple material options are currently under study. The plan is to produce an edge segment prototype (~1 m in length) and subject it to a full battery of tests. Testing will include: RoC measurement, light scattering properties, bending strain, and thermal strain. In addition, the prototype edge segment will be installed on an existing petal prototype (see Figure 6.2-1) to demonstrate the requisite installation precision.

Sharp optical edges also introduce a handling safety concern, partially mitigated by a nearly vertical bevel cut. The edge is generally blunt, except for the very small radius corner, and has limited capability to penetrate skin, when touched directly. Safety procedures will be strictly enforced and a remove-before-flight safety cover will be installed.

6.3.2   Starlight Diffraction Verification

Starshade optical performance will not be demonstrated by ground-based testing of any full-scale unit. The requisite distances are prohibitive. Rather, it will be demonstrated in a two-step process. First, subscale tests will demonstrate contrast performance consistent with imaging exo-Earths and validate the optical models, upon which full-scale shape tolerance allocations are based. The scaling approach is to match the flight design in terms of the number of Fresnel zones across the starshade, such that the diffraction equations defining the dark shadow are identical.

Second, shape tolerance allocations will be verified on the fully deployed flight unit. Key capabilities are already demonstrated via early prototypes (see Section 6.2). The status and plans toward the first step are detailed here.

6.3.2.1   Previous Test Results and Issues

Several experiments over the last decade demonstrate the viability of creating a dark shadow with a starshade, including: the University of Colorado (Schindhelm et al. 2007; Leviton et al. 2007); Northrop-Grumman (Samuele et al. 2009); Princeton University (Cady et al. 2009; Sirbu et al. 2013); and larger scale tests in a dry lake-bed (Glassman et al. 2013). Each of these experiments is limited in contrast performance to some extent by a subset of the following issues:

- Wavefront errors due to collimating optics
- Dust in open air testing
- Diffraction effects due to the finite extent of the optical enclosure
- Diffraction off starshade support struts

6.3.2.2   Current Test Results and Issues

The current starshade optical testbed at Princeton University addresses the limitations identified above to yield the darkest shadow produced by a starshade to date. An expanding beam is used to eliminate collimating optics and the corresponding contribution to Fresnel number is accounted for. It also helps to limit testbed length to a manageable level. Dust effects are limited by testing in an enclosed facility within an optical enclosure. Diffraction effects from the optical enclosure and support struts are mitigated with an innovative mounting scheme whereby the starshade is supported by an outer ring with an apodization profile optimized in similar fashion to the starshade profile. This introduces a non-flight outer working angle (OWA) limit at the tip of the outer ring.

Figure 6.3-4 shows the current starshade test article, as etched into a 4″ silicon wafer. The starshade has 16 petals with tips at an IWA of 400 mas. The optical edges are about 50 μm thick as compared to the 1 μm RoC specified for flight optical edges (see Section 6.3.1). The red circles are placed approximately at the IWA of 400 mas and OWA of 638 mas. The annular region between the IWA and OWA is the discovery space, where lies the dark hole. The OWA is unique to the test article and is not a feature of the flight starshade design.

Figure 6.3-4. Starshade test article supported by a diffraction controlling outer ring. Red circles indicate the inner and outer working angles.

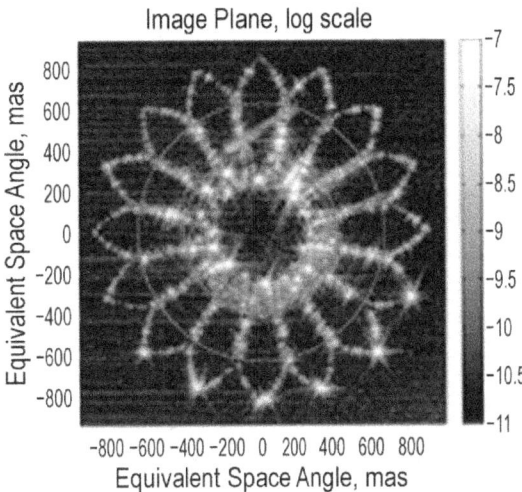

Figure 6.3-5. Measured contrast in image plane. Glint along the edge is greater than expected.

The starshade testbed at Princeton University is constrained to a 40′ long optical enclosure. A monochromatic laser operating at 632 nm simulates starlight. The testbed currently creates 590 Fresnel zones across the starshade, whereas the baseline flight design operates with 12 zones. Figure 6.3-5 shows the contrast measured in the image plane.

The edges are much brighter than expected and this is under investigation. One explanation is edge roughness and banding as result of the etching process. It is expected that the greater separation distance in the flight case will mitigate this problem. The power density of glint per unit length of edge would be the same, but would represent a much smaller fraction of the blocked starlight.

To obtain a quantitative measurement of the contrast achieved in the dark hole away from the glints, an azimuthal median was taken. A set of geometrical wedge constraints, as shown in Figure 6.3-6 were imposed to minimize the effect of the bright glint. Figure 6.3-7 compares the azimuthal median measurements to the diffraction theory predictions and glint, which is modeled here as point sources with matching peak intensity. The azimuthal median contrast measured in the intersection of annular and wedge regions is 2 to 3 orders of magnitude greater than diffraction theory predicts and is to large extent dominated by edge glint.

The median contrast across all 16 wedges at the IWA is about 1.0×10^{-10} and improves to about 2.5×10^{-11} at the OWA. The goal is to achieve contrast at 3×10^{-11} across the full annular region, as necessary to image exo-Earths after accounting for perturbations.

Not detailed here, for the sake of brevity, is the precursor testing of a circular-shaped control mask with the same outer ring configuration. This was used successfully to validate the calibration methodology and provide a reference point to compare the benefit of the optimized apodization profile.

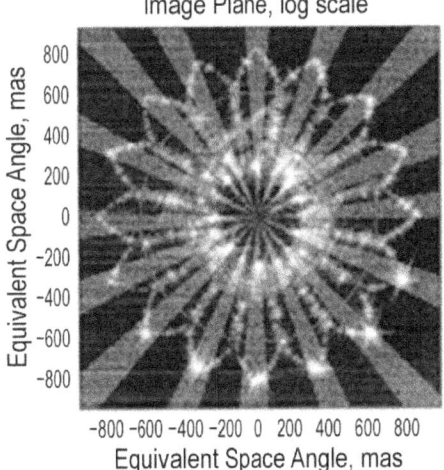

Figure 6.3-6. Wedge regions define azimuthal median contrast away from bright edges.

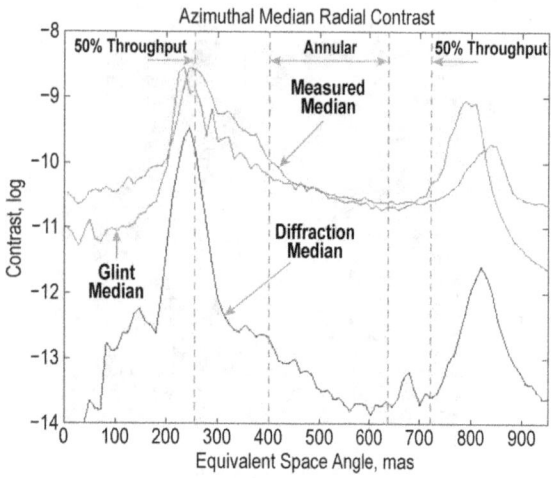

Figure 6.3-7. Azimuthal median comparison for the optimized starshade. The theoretical diffraction and glint simulations are shown and compared to the laboratory measurement.

6.3.2.3   Future Plans

A new TDEM activity (TDEM-12) led by Professor N. Jeremy Kasdin of Princeton University is underway to address optical performance verification and model validation. The development and testing of an improved subscale starshade with optical edge thickness ≤ 1 µm is the first priority. This is expected to greatly reduce starlight glint. An effort to model the observed glint is also planned. A completely new and much improved optical testbed is planned with length greater than 50 m. The goal is to achieve a Fresnel number within a factor of 2 to 3 of the baseline flight design. The starlight simulator will also be capable of producing broadband light.

A separate TDEM-12 activity led by Tiffany Glassman of NGAS is also underway to improve upon the open air testing of larger starshades, on the order of 1 m in diameter. Test objectives include characterizing sensitivity to lateral control errors and the benefit of spinning the starshade.

6.3.3   Formation Flying

Keeping the telescope within the dark-shadow created by the starshade (± 1 m control tolerance), at separation distances approaching 50,000 km, may seem like a daunting challenge. The overall formation flying design

is a challenging engineering problem that warrants further study in pre-Phase A. A specific aspect of the problem to qualify as an "unresolved technology issue" is formation sensing. Starshade lateral position must be sensed with 3-sigma accuracy better than ± 20 cm, relative to the telescope bore-sight, which is pointed at the target star.

Two factors make the formation sensing challenge tractable. First, the formation sensor utilizes the relatively large science telescope. It simultaneously images both the starshade laser beacon and the target star, as indicated by long, out-of-band wavelengths, that diffract into the shadow. Onboard image processing algorithms must estimate centroid positions with 3-sigma accuracy better than 0.3% of optical resolution. Built into these algorithms is a model of starshade diffraction at long wavelengths.

Second, the environment is very benign in Earth-leading orbit. Solar pressure is the dominant disturbance and yields a very low control bandwidth. This helps to keep formation control off the list of "unresolved technology issues", but also contributes to improving formation-sensing accuracy with long sensor integration times.

A new TDEM activity (TDEM-13) led by Professor N. Jeremy Kasdin of Princeton University is proposed to demonstrate the requisite formation sensing capability. A breadboard instrument, including FGS and image processing algorithms, will be integrated into the Princeton starshade optical testbed, as discussed in Section 6.3.2. The detector will be mounted on a 2-axis stage to simulate lateral position errors.

The proposed TDEM activity will develop the system design for formation flying and prototype algorithms for formation sensing, as discussed, in addition to trajectory estimator and formation control algorithms. Early simulations will demonstrate performance and assist in exploring optimal formation control and acquisition strategies, in terms of fuel usage. After integrating the instrument breadboard into the starshade optical testbed the control loop will be demonstrated with detector position stages simulating thrusters.

6.4 Summary

In summary, the Exo-S requirements are well understood. New technology development is focused on the starshade and formation sensing with manageable development risk. Several key on-orbit shape tolerance capabilities are already demonstrated with large margins:

- TDEM-09 demonstrated a manufactured petal width profile tolerance ≤ 100 µm.
- TDEM-10 demonstrated deployed petal root position random tolerances ≤ 0.5 mm (radial and tangential) and radial bias ≤ 0.25 mm.

Plans are in place to address the remaining open technology issues:

- TDEM-12 (Casement) will demonstrate optic edge radius ≤ 1 µm and reflectivity $\leq 10\%$ and stow radius at 1.5 m.
- TDEM-12 (Kasdin) will demonstrate scalable contrast performance to 3×10^{-11} and validate the starlight diffraction model.
- TDEM-13 (Kasdin) is proposed to demonstrate a lateral formation sensing tolerance ≤ 20 cm and validate starlight diffraction models in the FGS band.

Additional activities are proposed to mature technology readiness to TRL 5 prior to fiscal year 2017, but not detailed here. This includes repeating key performance demonstrations with higher fidelity prototypes.

7   References

Executive Summary

National Research Council (NRC) 2010, *New Worlds, New Horizons in Astronomy and Astrophysics* (Washington, DC: The National Academies Press).

Section 1

Bastien, F.A. et al. 2014, *AJ*, **147**, 29.

Batalha, N.M. et al. 2011, *ApJ*, **729**, 27.

Casertano, S. et al. 2008, *A&A*, **482**, 699.

Cassan, A. et al. 2012, *Nature*, **481**, 167.

Cash, W. 2006, *Nature*, **442**, 51.

Copi, C.J. and G.D. Starkman 2000, *ApJ*, **532**, 581.

Deming, D. et al. 2013, *ApJ*, **774**, 95.

Domagal-Goldman, S.D. et al. 2011, *Astrobiology*, **11**, 5.

Doyle, L.R. et al. 2011, *Science*, **333**, 6049.

Fressin, F. et al. 2013, *ApJ*, **766**, 81.

Guyon, O. and F. Martinache 2013, *BAAS*, **221**, #419.05.

Howard, A.W. 2013, *Science*, **340**, 572.

Kaltenegger, L. and W.A. Traub 2009, *ApJ*, **698**, 519.

Kopparapu, R.K. et al. 2013, *ApJ*, **765**, 131.

Lawson, P. 2013, *Exoplanet Exploration Program Technology Plan*, Appendix: 2012, http://exep.jpl.nasa.gov/files/exep/2012Appendix_Fall.pdf.

Lissauer, J. et al. 2014, *ApJ*, **784**, 44.

Madhusudhan, N. et al. 2011, *Nature*, **469**, 7328

Madhusudhan, N. et al. 2014, *Protostars and Planets VI*, ed. (University of Arizona Press), H. Beuther, R. Klessen, C. Dullemond, Th. Henning, submitted.

Marchal, C. 1985, *A&A*, **12**, 193.

Marley, M.S. et al., 2007, *ApJ*, **655**, 541.

Marois, C. et al. 2010, *Nature*, **468**, 1080.

Oppenheimer, B. and S. Hinkley 2009, *ARAA*, **47**, 253.

Pont, F. et al. 2013, *MNRAS*, **432**, 2917.

Petigura, E.A. et al. 2013, *Measuring Areas of Curves* (Proc. PNAS 110), ed. I.M.

Verma (PNAS: Washington, DC), 19273.

Rauer, H. et al. 2013, *Experimental Astronomy*, submitted (arXiv1310.0696).

Rogers, L.A. and S. Seager 2010, *ApJ*, **716**, 1208.

Rowe, J.F. et al. 2014, *ApJ*, **784**, 44.

Seager, S. 2013, *Science*, 340, 577.

Seager, S. et al. 2013, *ApJ*, **777**, 95.

Seager, S. and D. Deming 2010, *ARAA*, **48**, 631.

Schultz, A.B. et al. 2003, *High-Contrast Imaging for Exo-Planet Detection* (Proc. SPIE 4860), ed. A.B. Schultz and R.G. Lyon (Bellingham, WA: SPIE), 54.

Simmons, W.L. et al. 2004, *Optical, Infrared, and Millimeter Space Telescopes* (Proc. SPIE 5487), ed. A. Stohr, D. Jager, and S. Iezekiel (Bellingham, WA: SPIE), 1634.

Simmons, W.L. 2005, *A pinspeck camera for exo-planet spectroscopy*, Technical report, M.S. Thesis, Department of Mechanical and Aerospace Engineering, Princeton University.

Smith, M.W. et al. 2010, *Space Telescopes and Instrumentation 2010: Optical, Infrared, and Millimeter Wave* (Proc. SPIE 7731), ed. J.M. Oschmann, Jr.; M.C. Clampin; H.A. MacEwen (Bellingham, WA: SPIE), 773127.

Spitzer, L. 1962, *American Scientist*, **50**, 473.

Stapelfeldt, K. 2006, *Proceedings of IAU Symposium*, **232**, 149.

Sumi, T. et al. 2010, *ApJ*, **710**, 1641.

TESS, Website, "Overview", http://space.mit.edu/TESS/TESS/TESS_Overview.html

Vanderbei, R.J. et al. 2003, *ApJ*, **599**, 686.

Vanderbei, R.J. et al. 2007, *ApJ*, **665**, 794.

Wakeford, H.R. et al. 2013, *MNRAS*, **435**, 3481.

Woodcock, G.R. 1974, "Concept analysis and discussion: observations of extrasolar planets with an LST," Technical Report NAS9-14323, D180-18501-2, National Aeronautics and Space Administration, in *Future Space Transportation Systems Analysis*.

Zsom, A. et al. 2013, *ApJ*, **778**, 109.

Section 2

Backman, D. et al. 2009, *ApJ*, **690**, 1522.

Benneke, B. and S. Seager 2012, *ApJ*, **753**, 100.

Binney, J. and M. Merrifield, *Galactic Astronomy*, Princeton University Press, 1998, First edition.

Cahoy, K.L. et al. 2010, *ApJ*, **724**, 189.

Chiang, E. et al. 2009 *ApJ*, **693**, 734.

Coe, D. et al. 2006, *AJ*, **132**, 926.

Dawson, R.I. et al. 2011, *ApJ*, **743**, L17.

di Folco, E. et al. 2007, *A&A*, **475**, 243.

Eiroa, C. et al. 2013, *A&A*, **555**, A11.

Greaves, J. et al. 2004, *MNRAS*, **351**, L54.

Hinz, P. 2013, *AAS*, AAS Meeting 221, 403.06.

Illingworth, G.D. et al. 2013, *ApJS*, **209**, 6.

Karkoschka, E. 1994, *Icarus*, **111**, 174.

Kelsall, T. et al. 1998, *ApJ*, **508**, 44.

Koekemoer, A.M. et al. 2013, *ApJ*, **622**, 319.

Kuchner, M.J. and C.S. Stark 2010, *AJ*, **140**, 1007.

Kuchner, M.J. and M.J. Holman 2003, *ApJ*, **588**, 1110.

Kuhn, J.R. and S.L. Hawley 1999, *PASP*, **111**, 601.

Lee, J-M. et al. 2013, *ApJ*, **778**, 97.

Line, M.R. et al. 2014, *ApJ*, **783**, 70.

Nesvold, E. et al. 2013, *ApJ*, **777**, 144.

Nesvorny, D. et al. 2010, *ApJ*, **713**, 816.

Pirzkal, N. et al. 2005, *ApJ*, **622**, 319.

Roberge, A. et al. 2012, *PASP*, **124**, 799.

Robinson, T.L. et al. 2011, *Astrobiology*, 11(5): 393–408.

Rogers, L.A. and S. Seager 2010, *ApJ*, **716**, 1208.

Spyak, P.R. and W.L. Wolfe 1992, *Optical Engineering*, **31**, 1775.

Stark, C. and M. Kuchner 2008, *ApJ*, **686**, 637.

Windhorst, R.A. et al. 2011, *ApJS*, **193**, 27.

Virtual Planet Laboratory (VPL), Website, http://depts.washington.edu/naivpl/content/welcome-virtual-planetary-laboratory.

Section 3

Brown, R.A. 2005, *ApJ*, **624**, 1010.

Brown, R.A. et al. 2006, *Final Report of an Instrument Concept Study for a Wide-Field Camera for TPF-C*, Appendix C, www.stsci.edu/~RAB/papers/Mag_30_Cam.pdf.

Endicott, J. et al. 2012, *High Energy, Optical, and Infrared Detectors for Astronomy V* (Proc. SPIE 8453), ed. A.D. Holland and J.W. Beletic (Bellingham, WA: SPIE), 845304.

Section 4

Savransky, D. et al. 2010, *Space Telescopes and Instrumentation 2010: Optical, Infrared, and Millimeter Wave* (Proc. SPIE 7731), ed. J.M. Oschmann, Jr.; M.C. Clampin; H.A. MacEwen (Bellingham, WA: SPIE), 77312H.

Vanderbei, R.J. et al. 2007, *ApJ*, **665**, 794.

Section 5

Luedtke, A. and H.P. Stahl 2012, *Opt. Eng.*, **51**(5), 059701 (http://dx.doi.org/10.1117/1.OE.51.5.059701).

Scharf et al. 2010, *IEEE Systems Journal*, **4**(1), 84, Parts 1 and 2.

Stahl, H.P. et al. 2013, *Opt. Eng.*, **52**(9), 091805 (http://dx.doi.org/10.1117/1.OE.52.9.091805).

Section 6

Arenberg, J. et al. 2007, *Techniques and Instrumentation for Detection of Exoplanets III* (Proc. SPIE 6693), ed. D.R. Coulter (Bellingham, WA: SPIE), 669302.

Cady, E. et al. 2009, *Techniques and Instrumentation for Detection of Exoplanets IV* (Proc. SPIE 7440), ed. S.B. Shaklan (Bellingham, WA: SPIE), 744006.

Glassman, T. et al. 2010, *Space Telescopes and Instrumentation 2010: Optical, Infrared, and Millimeter Wave* (Proc. SPIE 7731), ed. J.M. Oschmann, Jr.; M.C. Clampin; H.A. MacEwen (Bellingham, WA: SPIE), 773150.

Glassman, T. et al. 2013, *Techniques and Instrumentation for Detection of Exoplanets VI* (Proc. SPIE 8864), ed. S.B. Shaklan (Bellingham, WA: SPIE), 886418.

Kasdin, N.J. et al. 2012, *Space Telescopes and Instrumentation 2012: Optical, Infrared, and Millimeter Wave* (Proc. SPIE 8442), ed. M.C. Clampin, G.G. Fazio, H.A. MacEwen, J.M. Oschmann (Bellingham, WA: SPIE), 84420A.

Kasdin, N.J. et al. 2013, *Techniques and Instrumentation for Detection of Exoplanets VI* (Proc. SPIE 8864), ed. S.B. Shaklan (Bellingham, WA: SPIE), 886417.

Leviton, D. et al. 2007, *UV/Optical/IR Space Telescopes: Innovative Technologies and Concepts III* (Proc. SPIE 6687), ed. H.A. MacEwen and J.B. Breckinridge (Bellingham, WA: SPIE), 66871B.

Martin, S.R. et al. 2013, *Techniques and Instrumentation for Detection of Exoplanets VI* (Proc. SPIE 8864), ed. S.B. Shaklan (Bellingham, WA: SPIE), 88641A.

Samuele, R. et al. 2009, *Techniques and Instrumentation for Detection of Exoplanets IV* (Proc. SPIE 7440), ed. S.B. Shaklan (Bellingham, WA: SPIE), 744004.

Schindhelm E. et al. 2007, *Techniques and Instrum7entation for Detection of Exoplanets III* (Proc. SPIE 6693), ed. D.R. Coulter (Bellingham, WA: SPIE), 669305.

Sirbu, D. et al. 2013, *OpEx,* **21**, 32234.

Shaklan, S. et al. 2010, *Space Telescopes and Instrumentation 2010: Optical, Infrared, and Millimeter Wave* (Proc. SPIE 7731), ed. J.M. Oschmann, Jr.; M.C. Clampin; H.A. MacEwen (Bellingham, WA: SPIE), 77312G.

Shaklan, S. et al. 2011, *Techniques and Instrumentation for Detection of Exoplanets V* (Proc. SPIE 8151), ed. S. Shaklan (Bellingham, WA: SPIE), 815113.

8   Acronyms

A&A	Astronomy & Astrophysics Journal	ESA	European Space Agency
ACS	Advanced Camera for Surveys	ESPRESSO	Echelle SPectrograph for Rocky Exoplanet and Stable Spectroscopic Observations
AJ	Astronomical Journal		
ApJ	Astrophysical Journal	ExEP	Exoplanet Exploration Program
ApJS	Astrophysical Journal Supplement	Exo-C	Exo-Coronograph
		Exo-S	Exo-Starshade
AFTA	Astrophysics Focused Telescope Asset	FF	formation flying
		FGC	fine guidance camera
ALMA	Atacama Large Millimeter/ submillimeter Array	FGS	fine guidance sensor
		FOV	field of view
ATLO	assembly, test, and launch operations	FRR	Flight Readiness Review
		FS	flight system
AU	astronomical unit	FSM	fast-steering mirror
BOE	basis of estimate	FSW	flight software
BOSS	Big Occulting Steerable Satellite	GMT	Giant Magellan Telescope
C/O	carbon to oxygen	GN&C	guidance, navigation, and control
C&DH	command and data handling	GPI	Gemini Planet Imager
CADRe	Cost Analysis Data Requirements	Gyr	gigayear
CAST	Control Analysis Simulation Testbed	HGA	high-gain antenna
		HOSTS	Hunt for Observable Signatures of Terrestrial planetary Systems
CATE	Cost Appraisal and Technical Evaluation		
		HST	Hubble Space Telescope
CCD	charge coupled device	HUDF	Hubble Ultra-Deep Field
CDR	Critical Design Review	HZ	habitable zone
CMM	coordinate measuring machine	I&T	integration and test
CoRoT	COnvection ROtation et Transits	IR	infrared
COTS	commercial, off-the-shelf	IWA	inner working angle
CTE	coefficient of thermal expansion	JPL	Jet Propulsion Laboratory
DOF	degree of freedom	JWST	James Webb Space Telescope
DRM	Design Reference Mission	KDP	Key Decision Point
DSMS	Deep Space Mission System	L	stellar luminosity
DSN	Deep Space Network	LBTI	Large Binocular Telescope Interferometer
DTE	direct to Earth		
E-ELT	European Extremely Large Telescope	limΔmag	planet contrast at the threshold of detectability
ELT	Extremely Large Telescope	LRR	Launch Readiness Review
EMC	electromagnetic compatibility	LV	launch vehicle
EMI	electromagnetic interference	MCR	Mission Concept Review
EOL	end of life		

MDR	Mission Definition Review	RoC	radius of curvature
MER	Mars Exploration Rover	RV	radial velocity
MIT	Massachusetts Institute of Technology	S/C	spacecraft
MNRAS	Monthly Notices of the Royal Astronomical Society	S/N	signal to noise
		SDC	Science Data Center
MOC	Mission Operations Center	SEP	solar electric propulsion
MRR	Mission Readiness Review	SMSR	Safety and Mission Success Review
MSL	Mars Science Laboratory	SNR	signal-to-noise ratio
NASA	National Aeronautics and Space Administration	SOC	Science Operations Center
		SPHERE	Spectro-Polarimetric High-contrast Exoplanet Research
NGAS	Northrop Grumman Aerospace Systems	SRR	System Requirements Review
NGC	Northrop Grumman Corporation	SSCM10	Small Satellite Cost Model 2010
NICM	NASA Instrument Cost Model	STDT	Science and Technology Definition Team
NICMOS	Near Infrared Camera and Multi-Object Spectrometer	STIS	Space Telescope Imaging Spectrograph
NIR	near-infrared		
NIRCam	Near Infrared Camera	STSci	Space Telescope Science Institute
NPR	NASA Procedural Requirements	TDEM	Technology Development for Exoplanet Missions
NRC	National Research Council		
NRO	National Reconnaissance Office	TESS	Transiting Exoplanet Survey Satellite
O^3	Occulting Ozone Observatory		
OPALS	Optical PAyload for Lasercomm Science	TMT	Thirty Meter Telescope
		TPF-I	Terrestrial Planet Finder Inteferometer
OpEx	Optics Express		
ORR	Operational Readiness Review	TRL	Technology Readiness Level
Osiris Rex	Origins Spectral Interpretation Resource Identification Security Regolith Explorer	TRR	Test Readiness Review
		UKST	UK Schmidt Telescope
PASP	Publications of the Astronomical Society of the Pacific	UMBRAS	Umbral Missions Blocking Radiating Astronomical Sources
		UV	ultraviolet
OWA	outer working angle	VLP	Virtual Planet Laboratory
PDR	Preliminary Design Review	VLT	Very Large Telescope
PLAR	Post Launch Assessment Review	WBS	Work Breakdown Structure
POSS	Palomar Observatory Sky Survey	WFIRST	Wide-Field Infrared Survey Telescope
PROBA	PRoject for OnBoard Autonomy		
PSF	point spread function	WISE	Wide-field Infrared Survey Explorer
QE	quantum efficiency		
R	reflectivity	WPF3	Wide Field Camera 3
R	spectral resolution	XDF	eXtreme Deep Field
RF	radio frequency		

www.ingramcontent.com/pod-product-compliance
Lightning Source LLC
Chambersburg PA
CBHW080600180526
45168CB00007B/2726